Business, Economics, and Law

Edited by
S. Zeranski, Wolfenbüttel, Germany
S. Reuse, Essen, Germany

In a knowledge-based scientific society it is necessary to fix excellent academic results as soon as possible in order to generate an interlink to practice. The academic series "Business, Economics, and Law" deals with innovative research results in business, economics, and law and promotes the dialogue between theory and practice. The series offers concrete impulses for further academic research and practical implementation.

Springer Gabler Results richtet sich an Autoren, die ihre fachliche Expertise in konzentrierter Form präsentieren möchten. Externe Begutachtungsverfahren sichern die Qualität. Die kompakte Darstellung auf maximal 120 Seiten bringt ausgezeichnete Forschungsergebnisse „auf den Punkt". Springer Gabler Results ist als Teilprogramm des Bereichs Springer Gabler Research besonders auch für die digitale Nutzung von Wissen konzipiert. Zielgruppe sind (Nachwuchs-)Wissenschaftler, Fach- und Führungskräfte.

Edited by
Stefan Zeranski
Brunswick European Law School
(BELS), Wolfenbüttel

Svend Reuse
FOM dips – Deutsches Institut für
Portfolio-Strategien, Essen

Alexander Zureck

Financial Communication in Small and Medium-Sized Enterprises

Patents in Financial Communication

With a foreword by Prof. Dr. Stefan Heinemann

 **Springer** Gabler

Alexander Zureck
Essen, Germany

ISBN 978-3-658-07486-9   ISBN 978-3-658-07487-6 (eBook)
DOI 10.1007/978-3-658-07487-6

The Deutsche Nationalbibliothek lists this publication in the Deutsche Nationalbibliografie; detailed bibliographic data are available in the Internet at http://dnb.d-nb.de.

Library of Congress Control Number: 2014952187

Springer Gabler
© Springer Fachmedien Wiesbaden 2015

Printed on acid-free paper

Springer Gabler is a brand of Springer DE.
Springer DE is part of Springer Science+Business Media.
www.springer-gabler.de

Foreword

Financial communication is becoming more relevant. Traditionally, financial communication was a relevant issue for listed companies. Today, also small and medium-sized enterprises have to master the challenges of financial communication. It addresses investors but also other stakeholders.

Thompson defines investor relations as "a strategic management responsibility that integrates finance, communication, marketing and securities laws compliance to enable the most effective two-way communication between a company, the financial community, and other constituencies, which ultimately contributes to a company's securities achieving fair valuation."(Adopted by the NIRI Board of Directors, March 2003) Therefore, efficient and effective strategies, tactics, and tools are necessary to meet stakeholders' needs.

Around the world, there is a trend to market financing. In contrast, traditional bank financing deemphasizes. This leads to the point that all companies have to deal with new financial partners. The traditional dialogue between management and bank is no longer up to date. Companies have to deal with different stakeholders from the financial community within the capital markets.

Especially for small and medium-sized companies it is a difficult task to shape new communication strategies to reach and bind investors. Because of the dominant bank financing with its different approaches in this sector in the past, these companies have a lack of experiences in the field of financial communication.

In Germany, initiatives like "Beste Finanzkommunikation für den Mittelstand" support small and medium-sized enterprises to improve their financial communication. Especially, the presenting of good examples like PFK Group GmbH or KAIMANN GmbH makes other companies think about their own strategies.

Alexander Zureck presents several aspects, how small and medium-sized enterprises can set themselves apart from the competition on the capital markets. He demonstrates that an improved financial communication improves a company's development in general.

The empirical study gives examples for companies from different business sectors, how they can improve their financial communication through implementing of intangible assets in their communication strategies.

The work combines the common literature and extends it through new findings. In general, the findings are impressing and they can improve companies' standing in competition. The findings are also useful for practitioners.

In sum: A valuable contribution for theory and practice.

Prof. Dr. Stefan Heinemann
Pro-Rector for Cooperations
FOM Hochschule

Preface

The MBA supports me in new ways of problem solving. In the past, I tried to solve problems without considering all relevant aspects. My MBA studies help me to analyze problems from different points of view. Especially, the dialogue with my colleagues from other scientific disciplines in the MBA improved my own capabilities to analyze and solve problems.

It was important for me to deal with an interdisciplinary topic. With regard to the fields of my MBA studies and my job, I wanted to concentrate on a topic dealing with financial and communicational issues. With the chosen topic, I tried to connect both topics. Also, I wanted to give a solution to currently discussed problems in companies' practice.

Paying particular attention to the fact that I did my MBA in part-time, I have to thank several persons who supported me during the time of writing the MBA dissertation.

First of all, I would like to thank my family and my friends. It took a lot of time to get the MBA. So I had to neglect some social contacts as well as some activities in leisure time. I would like to address special thanks to my mother Susanne for supporting me during my former education and while I was preparing the MBA dissertation. Just the same, I have to thank my girlfriend Fabienne and my father Ralf.

Moreover, I have to thank Prof. Dr. Stefan Heinemann, who accompanied my works on the MBA dissertation with a lot of engagement as well as a lot of useful ideas. It was nice to discuss all relevant aspects.

Certainly, I have to thank Prof. Dr. Eric Frère for his support during my MBA studies as well as in my former studies and in the job. In times of pressure, he guided me to good results. Furthermore, he enables me to continue my scientific way.

Further, my thanks go to Prof. Dr. Julius Reiter. Dozens of times, he debated with me on financial issues from a lawyer's point of view. The debates opened my way of thinking.

Furthermore, I have to thank Dr. Svend Reuse for the ongoing professional input to my works.

Moreover, I have to thank my employer FOM Hochschule for supporting me during the MBA.

Finally, the greatest thanks go to my friend and colleague Tino Bensch. Besides correcting my master dissertation, he provided a lot of professional ideas to improve my work.

Alexander Zureck

Executive Summary

The financial crisis has confirmed that there are no perfect markets. The financial markets are almost perfect, but at the end they are not really perfect. Moreover, there are information asymmetries in the financial markets. It is important to handle and to minimize these information asymmetries.

To ensure refinancing opportunities for small and medium-sized enterprises, answers for bridging the information gap between investor and company are needed. In this context, the capital markets become increasingly important. Classical bank financing becomes more difficult as a result of new regulatory requirements of the future. Traditionally, banks fulfill the risk transformation function. If a company uses the capital market for refinancing, the market performs this function. That is the reason why companies have to adapt their refinancing practice to capital market requirements. The market has to be able to evaluate the risk for investing in the company. Companies have to check out the requirements of the capital market to guarantee their refinancing possibilities in the long run. Therefore, they have to communicate the capital markets demanded by the market.

This work focuses on patents as an example of intangible assets and on their importance for the financial communication between small and medium-sized enterprises and the capital market owing to closing the information gap between company and investor. A positive correlation between patent application and stock price development underscores the importance of patents in financial communication and as a mean of closing the gap.

The research within this work shows a positive correlation between patent application and share price performance of small and medium-sized enterprises. The findings lead to recommendation, taking patents and other intangible assets into company's financial communication.

In addition to the general importance of patents in financial communication, some companies have a disproportionate significance of patents in financial communication. This work illustrates that industry sectors with short-term and very innovative products as well as those with long-term and very cost-intensive products profit strongly from patents in financial communication. This point can be confirmed also for companies with a small number of employees. More often, these companies use patents on their own in comparison to bigger companies.

Generally speaking, more than 65% of all investigated patents are significant. The 173 significant patents include 468 significant days or investigation periods. 161 of 468 patents are significant on 95% level of significance and 130 patents of 468 on a 99% level of significance. Therefore, more than 62% are significant on a level of significance equal to or more than 95%.

Due to these empirical findings, companies fulfilling the following parameters profit more than other companies from patents in their financial communication:

80% of all patents from companies from the energy sector, 77% from the communications sector, and 75% from the consumer sector have significant impacts on companies' share prices,

69% of all patents from companies with less than 150 employees have significant impacts on companies' share prices.

At the end, small and medium-sized companies have to implement patents and other intangible assets into financial communication for mastering the war for capital. Each company has to find its role in the competition for capital. A number of companies need money, but their investors' means of capital are limited. Therefore, companies have to demonstrate their capabilities not only to customers but also to investors. A communication strategy targeting all stakeholders is very important for success in the future.

Table of Contents

List of Abbreviations

art.	article
BGBl	Bundesgesetzblatt
BIS	Basel Committee on Banking Supervision
CAPM	Capital Asset Pricing Model
cp.	compare
e.g.	for example
ed.	editor
edn.	edition
eds.	editors
EPÜ	Europäisches Patentübereinkommen
et al.	and others
etc.	et cetera
EU	European Union
EUR	Euro
ff.	following pages
IFRS	International Financial Reporting Standards
IPC	International Patent Classification
Iss.	issue
KfW	Kreditanstalt für Wiederaufbau
Mio.	million
NIE	New Institutional Economics
No.	number
para	paragraph
PatG	Patentgesetz (patent law)
pp.	pages
R&D	Research and Development
SME	Small and Medium-Sized Enterprises
URL	Uniform Resource Locator
US-GAAP	United States Generally Accepted Accounting Principles

USP	Unique Selling Proposition
Vol.	volume
w/o	without

List of Figures

List of Tables

List of Symbols

&	and
$\geq$	greater/equal than
§	paragraph
§§	paragraphs
%	percentage
$\leq$	smaller/equal than

List of Formulas

List of Formulas

1   Introduction

The present work starts with an outline of the further content of this work. It is based on the origin of the problem in the financial crisis. Stemming from the financial crisis, new possibilities for handling the problem and the way to check these possibilities in the further work are described in Chapter 1.

1.1   Problem Definition and Objective

The last financial crisis has far-reaching impacts on the financial sector. The financial sector changes from the practical as well as from the theoretical perspective. In general, there is a stronger focus on economic certainty. Thus, it is possible to mention Basel III and other new requirements because of greater financial and economic stability. Furthermore, there are changes in the theoretical base of financial research.[1]

Until such time as the crisis erupted, most research in the field of finance was based on the assumption of perfect markets. Since the last crisis, research has not perpetuated all given assumptions in connection to perfect markets. The information asymmetry between all market participants has become all the more important for financial research.

There are different information asymmetries. In the following section, the information asymmetry between the issuer of a security and the investor, the shareholder, is considered in detail. A strong relationship between these protagonists is relevant for a stable and functioning economic cycle.

Changes in the field of regulation, e.g. Basel III, require adjustments by companies due to the acting in field of refinancing. Companies have to fulfill a lot of requirements to receive fresh money from traditional lenders like banks. Nowadays, banks often have strict lending policies. As a result, companies look closer to alternative sources to refinance their businesses.

The capital market becomes more important in comparison with the traditional bank financing. Traditionally, banks fulfill the risk transformation function. If a company directly operates at the capital market, the capital market fulfills the

---

[1]   Cp. Duca, J., Muellbauer, J., Murphy, A. (2010), pp. 203-217.

risk transformation. In this context, market participants have to be able to evaluate the risks and chances that occur.

These risks and chances are meant to evaluate information with regard to one company, on a micro level, and to the whole economic system, on a macro level. From the issuer's and investor's point of view, both levels are important for decision-making. Especially, the information on micro level is relevant because it influences once stock price disproportionately.

Patents are an example of intangible assets. It is more difficult to identify the real value of an intangible asset than of a tangible asset. The following work looks at the relevance of patents in financial communication and the impacts on company's stock price performance. In contrast to the past, today intangible assets are very important in several industry sectors such as the biotech and the information technology sector. The number of tangible assets in these sectors is rare.

The work seeks to find a positive correlation between patent application and stock price development. It is important to stress a link between intangible assets and an increasing corporate value in general. Hence, the general impact of patent applications on the stock price of a company is analyzed in the beginning. Thereupon, the analysis is differentiated due to a company's sales, number of employees, industry sector, the IPC class of patents, and the publication year.

1.2   Scope of Work

The scope of work is to analyze the role of patents and other intellectual properties in the financial communication of SMEs. To answer this question, the work is divided into five chapters.

Based on the theoretical background of the problem, the work gives a more and more closer look at the problem. The empirical analysis in Chapter 5 analyzes the problem for a much delimited choice of companies. The following graphic gives an overview of the approach used in the present work for analyzing the problem:

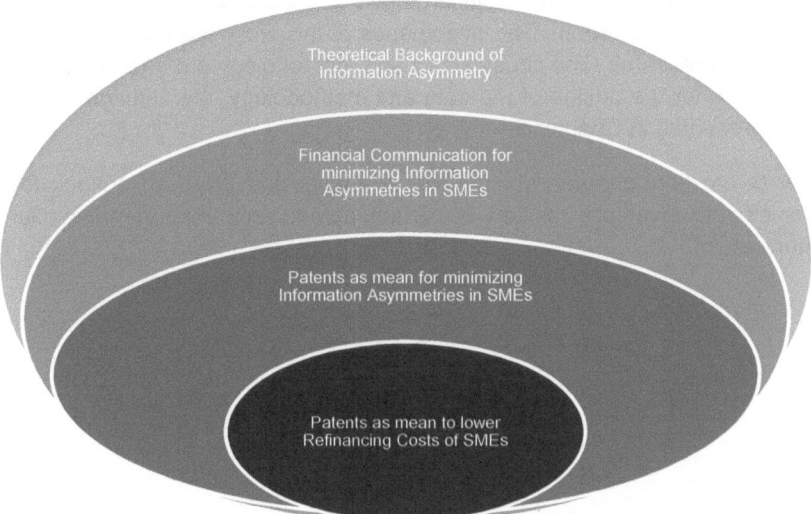

Figure 1: Methodical Approach of the Work
Own graphic.

Chapter 2 gives the theoretical background to the topic. In this chapter, the financial communication within the information economics is described. Therefore, there is an introduction to New Institutional Economics and the principal-agent problem at the beginning. It is followed by a link between theory and practice with a focus on financial communication.

Chapter 3 makes the financial communication a subject of the German SMEs. After that, German SMEs are analyzed. It is followed by a closer look at the general financing situation of these companies. The chapter ends with a discussion about the financial communication partners of SMEs vis-à-vis other companies, especially big companies.

Chapter 4 outlines the importance of patents as a kind of intellectual property within the financial communication. For this, firstly, there is a presentation of the different patent classifications, which is followed by the process of patent application. The chapter ends with a discussion of the secondary literature on the role of patents within financial communication.

The fifth chapter includes the empirical analysis. It starts with an overview of the current state of research. The hypotheses of this work are outlined based on the state of research. The chosen data and methodology are oriented to the hypotheses. After the outline of the data and methodology, the findings of this work are presented in Chapter 5.

Chapter 6, the final chapter, summarizes the work and gives an outlook for companies and investors on how important patents are in further financial communication.

2 Financial Communication in the Context of Information Economics

Chapter 2 is the theoretical section of this work. Based on the problems described in Chapter 1, Chapter 2 stresses the theoretical background of the mentioned problems. Therefore, the chapter starts with a complete overview of the New Institutional Economics. Following this, there is a closer look at the principal-agent approach for analyzing the information asymmetry between management and shareholders.

2.1 Introduction to New Institutional Economics

New Institutional Economics (NIE) deals with institutional and organizational issues in economics. The way to deal with these problems distinguishes in style, methodology, and contents in the available literature. The literature started concentrating on these problems after World War II. Researchers recognized that institutions have an important and critical role in economic performance. Prior to World War II, the focus was on mathematical development of neoclassical theory. Researchers tried to develop more and more abstract economic models. Today, there are different approaches to the NIE: transaction-cost economics, property rights analysis, legal economics, constitutional economics, public choice theory, the principal-agent approach, relational contracts, comparative economic systems, etc.[2]

In contrast to the neoclassical theory, the NIE analyzes company's role in the economy. The neoclassical theory only touches on different institutions but not their role in the general economy. The NIE is a modified theory of the existing neoclassical theory. There are differences in several ways. However, there are also similarities because the NIE uses standard economic theory for analyzing the functioning of institutions. The analysis is important to evaluate institutions' role in economic operations.[3]

The technological development in economy has an important role in institutional research. Institutional research does not focus on historical-deductive studies, the object of research is on institutional framework, and the implica-

---

[2] Cp. Furubotn, E., Richter, R. (2008), p. 15.
[3] Cp. Coase, R. (1984), p. 230.

tions of given institutional arrangements for economic behavior. Less focus is paid on spontaneous social organizations; more attention is on constructed organization developed with technological support.[4]

One main part of research in the context of the NIE lies in transaction costs. Transaction costs are a consequence of all doings of the protagonists involved in an economic system. Operational economic systems produce a lot of transactions costs for all participants.[5]

The NIE is based on the idea that economic activity takes place in neoinstitutional system. There are positive transaction costs in this system and the decision-makers are rationally bounded. Furthermore, all exchanges of goods and other things generate transaction costs. All market participants changing goods and other things are forestalled by this. They are not trustworthy and engage in opportunistic behavior.[6]

There are two researchers who give their approaches concerning the NIE. On the one hand, there is Williamson and on the other there is North. Williamson focuses on the institutional arrangements. These arrangements are, for example, a firm's organization or its contracts. In contrast to Williamson, North concentrates on the institutional environment of a social system. The environment can be the institutional framework. Williamson assumes an institutional environment.[7]

The researchers have different opinions regarding transaction costs and the design of institutions. Williamson mentions that institutions are designed to reduce transaction costs. Only formed institutions are able to survive in competitive markets. According to North, institutions exist to reduce transaction costs. In turn, they make the economy efficient. Inefficient markets exist because of a plethora of policy in the markets.[8]

In summary, the NIE has different forms which are driven by their issuers. In general, the NIE is a theory with a lack of formalism. However, the deep level

4   Cp. Hayek, F. (1973), w/o p.
5   Cp. Arrow, K. (1969), p. 48.
6   Cp. Coase, R. (1984), p. 231; North, D. (1995), pp. 18-19; Williamson, O. (1975), p. 4.
7   Cp. Davis, L., North, D. (1971), p. 133; Williamson, O. (2000), p. 597.
8   Cp. Ensminger, J. (1992), pp. 21-22; Furubotn, E., Richter, R. (2005), pp. 108-110; North, D. (1986), p. 236; North, D. (1990), p. 8 and 52; Williamson, O. (1981), pp. 1537-1568.

of formalism forces its role in literature. The NIE makes valuable contributions to several problems and it is suitable for different issues.[9]

While Williamson has further developed the transaction cost theory (1967 and 1975), Coase (1960), Furubotn and Pejovich (1972), as well as Alchian and Demsetz (1972 and 1973) focus on the idea of property rights. The definition and the way of distribution are interesting in this context. The transaction cost theory concentrates on a single transaction and the property rights theory considers the transfer of a bundle of rights attached to the physical commodity or service, since it is the value of the rights (and obligations) that determine the value of what is exchanged.[10]

Jensen and Meckling (1976), Fama and Jensen (1983a, 1983b), as well as Holmström (1979 and 1982) look from a third person's point of view at the form and function of an economic institution. For these researchers, the role of information is an object of research. In this context, they look closer at the relationship between the principal (i.e. the shareholder) and the agent (i.e. the management). Normally, the agent should act on behalf of the principal but often they are opponents with different information levels. Both are human beings with self-interest and the desire to maximize their self-interest. There is a conflict owing to their target function. This problem refers to the agency theory which is discussed in the next chapter.[11]

Firms, markets, and institutions have different low-rationality agents. The agents seek to reach their decisions with their own ways for decision-making. Everybody tries to optimize their own procedures. Focus lies on quick and cheap transactions between the opponents. The NIE explains limitations of time, resources, cognitive abilities, and possibilities to solve opponents' problems.[12]

The following illustration gives an overview of the NIE and its related theories:

---

[9]   Cp. Furubotn, E., Richter, R. (2008), p. 18.
[10] Cp. Alchian, A., Demsetz, H. (1972), pp. 777-795; Alchian, A., Demsetz, H. (1973), pp. 16-27; Coase, R. (1960), pp. 1-44; Demsetz, H. (1967), p. 347; Furubotn, E., Pejovich, P. (1972), pp. 1137-1162; Williamson, O. (1967), pp. 123-138; Williamson, O. (1975), w/o p.
[11] Cp. Fama, E., Jensen, M. (1983a), pp. 301-325; Fama, E., Jensen, M. (1983b), pp. 327-349; Holmström, B. (1979), pp. 74-91; Holmström, B. (1982), pp. 324-340; Jensen, M., Meckling, W. (1976), pp. 305-360; Kim, J., Mahony, J. (2005), p. 224.
[12] Cp. Conlisk, J. (1996), pp. 675-676; Furubotn, E., Richter, R. (2008), pp. 20-21; Gigerenzer, G., Selten, R. (2001), p. 14.

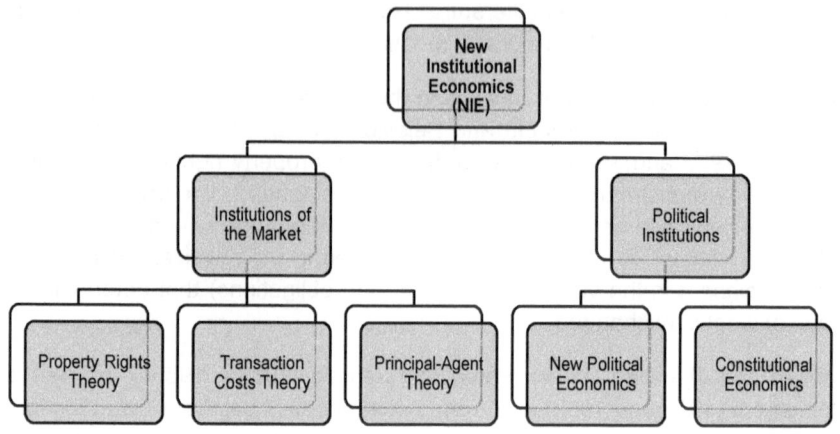

Figure 2: New Institutional Economics (NIE)

Own graphic according to: Coase, R. (1984), pp. 229-231; Demsetz, H. (1967), pp. 347-359 and Kim, J., Mahony, J. (2005), pp. 223-242.

2.2 Introduction in the Principal-Agent Problem

Basically, the principal-agent model has two different parties. These parties are the principal and the agent. Principal and agent have a contract. The agent performs some service on behalf of the principal. Therefore, the agent receives remuneration. Furthermore, the principal delegates some decision-making authority to the agent.[13]

In reality, there are a lot of principal-agent relationships. Both the principal and the agent have the choice to get into contract with other partners, which is why the principal has to offer an attractive compensation package for all tasks the agent does. On the other hand, the agent is limited in relation to their capabilities regarding the total remuneration.[14]

The following graphic shows the typical relationship between principal and agent:

---

[13]   Cp. Besley, T., Ghatak, M. (2005), p. 151; Jensen, M., Meckling, W. (1976), p. 308.
[14]   Cp. Besley, T., Ghatak, M. (2005), p. 151.

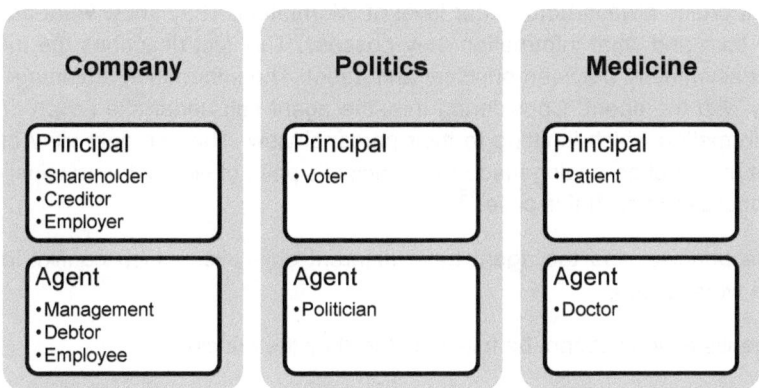

Figure 3: Principal-Agent Relationships
Own graphic according to: Besley, T., Ghatak, M. (2005), p.151; Erlei, M. et al. (2007), p.75;
Picot, A. et al. (2008), p.72.

In a company there are several relationships between principal and agent. In general, there is a network between individuals. These individuals can be a principal in one relationship and in another they are agent. The changing roles can be characterized as a chain of command and control. These functions change with regard to each role of the individual.[15]

The principal-agent theory focuses on tasks which the principal delegates to the agent. The reason for this is the principal who cannot execute these tasks on their own. These happen when a task requires some special expertise or the principal has limited amount of time.[16]

Delegation is important in the context of productivity and returns. As early as 1776, Adam Smith described the way firms raise productivity and returns with delegation of several tasks. These tasks can be in the nature of things the agent has to do or in the nature of reporting some piece of information that the principal needs for decision-making. Generally, all delegated tasks impact the principal's money. In this situation, the principal tries to push the agent in their work for maximizing the principal's welfare.[17]

---

[15]   Cp. Besley, T., Ghatak, M. (2005), p. 151.
[16]   Cp. Laffont, J., Martimort, D. (2002), p. 28.
[17]   Cp. Besley, T., Ghatak, M. (2005), p. 151; Jensen, M., Meckling, W. (1976), p. 309;
      Smith, A. (1776), pp. 2-6.

The agent profits always from better level of information. They know what action they take and what information they possess. This fact describes the information asymmetry between principal and agent. The information advantage is usually with the agent. Considering this, the agent can undertake action or report information which is fitting to their personal view. The personal view of the agent must not be analogous to the principal's point of view. The principal-agent model assumes in this case:[18]

1. The principal and the agent have different objectives which conflict in the worst cases,

2. Agent's actions cannot be fully monitored by the principal.

The agent always has private information. These pieces of information are not verifiable for the principal. The principal has no idea whether the agent acts in the principal's or in their own way. This situation is the starting point in the pursuit to increase the advantageousness of the principal-agent relationship.[19]

In theory, a world where all information is accessible free of charge, all economic actors are on the same level of information. In such a world, delegation increases company's structures for more productivity and more returns. All actions would be specifiable ex ante (in advance) and no one had the leeway in behavior to deviate from contractual agreements because the counterpart would know how to inhibit this behavior. Structures leading to a maximum of economic wealth under these conditions are called first-best solutions.[20]

In reality, companies are not able to get all information. If a company gets all information, it would be possible to allocate all resources in the best way. That is theory because the costs for gathering all information are high and they are not in relation to possible earnings. So, in reality, knowledge of economic actors remains incompletely and unevenly distributed. This situation creates an advantage for the agent. They get the possibility to act in their own interest because of less supervision by the principal. This could lead to negative repercussions for the principal. In practice, the principal reacts to this situation with limitation on the agent's leeway. For this purpose, the principal uses several additional means such as monitoring, control mechanisms, and appropriate incentives. Depending on how many means the principal uses, the advantage

---

[18] Cp. Besley, T., Ghatak, M. (2005), p. 151; Laffont, J., Martimort, D. (2002), p. 28; Picot, A. et al. (2008), p. 72-73.
[19] Cp. Besley, T., Ghatak, M. (2005), p. 152; Picot, A. et al. (2008), p. 72.
[20] Cp. Laffont, J., Martimort, D. (2002), p. 3; Picot, A. et al. (2008), p. 72.

of delegation including the earnings due to specialized knowledge of the agent could be minimized. In summary, the first-best solution cannot be reached in practice. Incomplete and asymmetric information lead always to the second-best solution. The difference in costs between the first-best and the second-best solutions lie in the agency costs.[21]

According to Jensen and Meckling, the agency costs are the sum of the following:[22]

1. **Monitoring expenditures:** Principal's costs for measuring and observing the agent's behavior. These include costs for efforts to control agent's behavior through budget restrictions, compensation policies, operating rules, etc.,

2. **Bonding expenditures:** Agent's costs to demonstrate that they will not injure the principal's interests,

3. **Residual loss:** The principal loses welfare despite monitoring and bonding expenditures. The agent does not get all relevant information for productivity-maximization (coordination problem) and this leads to opportunistic undermining of the structure (motivation problem). Residual loss is equivalent to the principal's reduction of welfare.

The agency costs include a trade-off. Increasing monitoring and bonding expenditures can reduce the residual loss. Generally, the agency costs are related to selected institutional design and a number of general institutional conditions. These, for example, include valid rules and norms that limit individuals' behavior. The normative approach of the principal-agent problem concentrates on finding a situational minimization of agency costs. Within this situational minimization, the second-best solution is closely connected to the first-best solution. For this, the agent needs appropriate monetary and non-monetary incentives. For measuring the institutional design or the contractual design as one part of the institutional design, the agency costs are the efficiency criteria for resource allocation.[23]

Besides the normative question, there is a positive question about the consequence of institutional arrangements. This one is less mathematical but more empirical. In the 1970s and 1980s, research focused on a special case of prin-

---

21   Cp. Erlei, M. et al. (2007), p. 75; Picot et al. 2008, pp. 72-73.
22   Cp. Jensen, M., Meckling, W. (1976), p. 309.
23   Cp. Jensen, M., Meckling, W. (1976), p. 309; Picot, A. et al. (2008), pp. 28, 36 and 72-73.

cipal-agent problem between owners and managers of large companies. Researchers and positive theorists concentrated on identifying conflicting situations between principal and agent. Based on these situations, they described governance mechanisms limiting the agent's self-serving behavior. Jensen and Meckling analyzed the relation between equity owners of a listed company and the confronting interests of employed managers. In this context, Fama focused on the varying interests of employees and employers. Both Fama and Jensen extended their research and concentrate on the different roles of shareholders, managers, and other employees within a corporate setup. In total, there is different research to solve existing agency problems within companies and within capital markets.[24]

Normative research and positive research differ in focus and style. But there is consensus in common assumptions about people, organizations, and information as well as a common unit of analysis. Research is based on institutional arrangements between principal and agent. Because of these commonalities, both approaches complement each other. The positive approach identifies various contract alternatives. The normative approach indicates the most efficient contract under certain conditions.[25]

Analyzing the conflict between principal and agent, the agent is prepared to take more risks than the principal. This risk deviation is important for further discussion within this work. The principal's and agent's behavior is related to their respective level of risk. They act according to their risks. Considering this, the principal-agent theory seeks to solve two problems that occur in an agency relationship—the agency problem (conflicting objective and monitoring difficulties/costs) and the problem of sharing risk. Therefore, economics of information and choice under uncertainty lead to the main variables worked within the principal-agent model: information, outcome uncertainty, and risk behavior.[26] Possibilities to react to these problems in financial context are discussed in the next chapter.

---

[24] Cp. Eisenhardt, K. (1989), pp. 59-60; Fama, E. (1980), pp. 208-307; Fama, E., Jensen, M. (1983a), pp. 301-325; Jensen, M. (1983), p. 334; Jensen, M., Meckling, W. (1976), p. 309.
[25] Cp. Eisenhardt, K. (1989), pp. 59-60; Jensen, M. (1983), p. 334.
[26] Cp. Eisenhardt, K. (1989), pp. 58-60.

2.3 Financial Communication as Consequence of Information Asymmetry

For decision-making, actors need information. Information is important to act in favor or against it. In the preceding chapter, the relationship between principal and agent is described. Both are on another level of information. One of them has more information than the other party. This difference in the level of information is called information asymmetry.[27]

The better informed party is the agent. There are different types of informational advantages:[28]

1. **Hidden characteristics:** The agent has private knowledge which is unknown to the principal before getting into a contractual relationship (ex ante). This knowledge mostly affects agent's costs and capabilities,

2. **Hidden information:** If principal and agent are in an existing agency relationship (ex post), the principal will have no knowledge about the amount of information available to the agent,

3. **Hidden action:** The principal does not know all actions taken by the agent in an existing agency relationship (ex post),

4. **Hidden intention:** The agent acts opportunistically. The principal does not know the agent's aims (ex ante and ex post).

In the financial sector, hidden characteristics often appear when a company initially uses the capital market for refinancing. The company has private information unknown to the market and its participants. This information can be related for further development of the company's business. The intensity of existing business practices with suppliers and customers is not assessable. Moreover, the unique selling proposition of the company is not analyzable because there is no comparison.

By investing money in a company, an investor enters into a contractual relationship with the company. An investor is normally not engaged in the company and therefore information is hidden. They just have a close look at the company and its processes as well as the whole business. In this role, the

---

[27] Cp. Besley, T., Ghatak, M. (2005), p. 151; Frank, R. (1994), p. 203; Laffont, J., Martimort, D. (2002), p. 2-5.
[28] Cp. Laffont, J., Martimort, D. (2002), p. 3; Picot, A. et al. (2008), p. 75.

company's management is the agent and it is well-informed to all points affecting the company.[29]

With regard to hidden information, there are hidden actions by the company's management. The investor does not know about all actions taken by the management. Furthermore, they cannot evaluate the actions which are closely related to hidden intention. Concretely, the investor cannot evaluate, for example, the benefit of a new service contract. They do not know the influences of new contracts on their dividend payout. There are a lot of processes and decisions made by management, which the investor does not get to know. All this information can be important for them but at the end, they have to trust the management.[30]

In connection with the information asymmetries, three different types of agency problems can be differentiated-adverse selection, moral hazard, and hold-up. These types exist in several principal-agent relationships.

Adverse selection

Adverse selection emerges ex ante. There, the principal does not know the hidden characteristics of the agent. At the same time, the agent does not know which characteristics they have to offer to the principal to get into contract. The principal decides for or against one agent with a limited amount of information. Primarily, if the agent is into a contract, the principal will get a wider view into the agent's real characteristics. Ex post the principal has more information to evaluate their decision. Agents with below-average or disadvantageous attributes can take this situation to get into contract although they have less capabilities for the task. Conversely, agents do not get into a contract although they have fitting attributes for the task. These agents hide their good attributes for the task during getting into contract with the principal. The principal decides against them because they do not gain an insight into agent's real characteristics.[31]

In the financial context, adverse selection emerges often. Especially big companies invest a lot of money into investor relations to beam attention toward the company. Smaller firms often do not pay attention to financial communica-

---

[29]   Cp. Ueda, M. (2004), pp. 601-621.
[30]   Cp. Kester, W. (1192), pp. 25-26.
[31]   Cp. Picot, A. et al. (2008), p. 74.

tion with potential investors. At the end, bigger companies do not have a better performance than the smaller ones. But the smaller ones are unknown to a plenty of investors.[32]

Moral hazard

In contrast to adverse selection, moral hazard emerges in the course of an existing principal-agent relationship. The principal delegates several tasks to the agent. The agent executes these tasks. Simultaneously, the principal has to monitor the agent. In many cases, the principal is not able to monitor agent's work. There are different reasons for this. One of the simplest ones is that the principal does not have the time to monitor the agent. That is a typical situation for a moral hazard within hidden action. Another situation describes moral hazard with hidden information as challenge. If the agent is a real specialist in their field of work, the principal will not be able to judge the agent's work because they are no expert in the same field of work.[33]

Both situations are difficult to evaluate for the principal. The principal receives an outcome upon finishing the task. Because of the hidden action and the hidden information, the principal cannot estimate the agent's contribution to the total outcome. He is not able to differentiate between agent's performance and other exogenous factors which push the total outcome.[34]

In the financial background, the management (agent) can reach a certain outcome using an unfavorable market situation just to override the bad results of previous easy-going mode of task performance (hidden action) or even attitude (hidden information). It is a dangerous situation for the shareholder (principal) because the agent capitalizes on the principal's information disadvantage in an opportunistic manner. This leads to moral hazard. The principal is not able to judge the agent's decisions.[35]

---

[32]   Cp. Grunig, J., Hunt, T. (1984), w/o p; Laskin, A. (2006), pp. 69-70.
[33]   Cp. Picot, A. et al. (2008), p. 75.
[34]   Cp. Erlei, M. et al. (2007), pp. 109-110.
[35]   Cp. Picot, A. et al. (2008), p. 75.

Hold-up

The hold-up is connected to hidden intention. The principal and the agent do not know each other's attitude and the motives for getting into a contract. Both are on a low level of information before entering into a contractual relationship. They do not know how the other party would act during the contractual relationship. Both have the chance to act in an opportunistic way to increase personal outcome. As a result, both observe each other during the contractual relationship.[36]

It becomes problematic when the principal gets into a situation where they become dependent on the agent. The principal depends often on the agent when they previously did high investments which are irreversible. In this case, the principal has to conclude that the agent works consistently in their interests. This conflict of interests can appear in all contractual relationships between principal and agent.[37]

In the area of finance, the hold-up often appears in connection with transaction costs. Investors have to pay transaction costs to get into a contract with a principal. For the investors, it is important that the transaction costs amortize in the future. Therefore, it is relevant that the management (agent) works in shareholder's (principal's) interests. This situation is a matter of analysis on specific investments as part of the previously mentioned transaction cost theory in context of NIE.[38]

For solving the problem of hold-up, efficient contracts are needed. Contracts have to include a number of provisions for possible eventualities. Just in this way, the danger of opportunistic behavior on the agent's site is on a low level. The agent acts in the principal's interest because they are rewarded for all good things.[39]

After introducing the different types of agency problems and their origin, the work now focuses on the behavioral control of interests between principal and agent. The general objective of all means of control is to reach an efficient use of economic resources. There are different solutions based on the interdependency of asymmetric information, risk allocation, and incentives. It is the aim to establish institutional settings between principal and agent that raise the

---

[36] Cp. Klein, B. et al. (1978), p. 297; Picot, A. et al. (2008), p. 75.
[37] Cp. Breid, V. (1995), p. 825; Picot, A. et al. (2008), p. 75.
[38] Cp. Picot, A. et al. (2008), pp. 59 and 76.
[39] Cp. Picot, A. et al. (2008), p. 75; Williamson, O. (1975), p. 32.

level of information. All means of control in the relationship between principal and agent is bound to welfare. It is always a compromise between the two. In general, there are different mechanisms to discipline the agent and to get the relationship under control.[40]

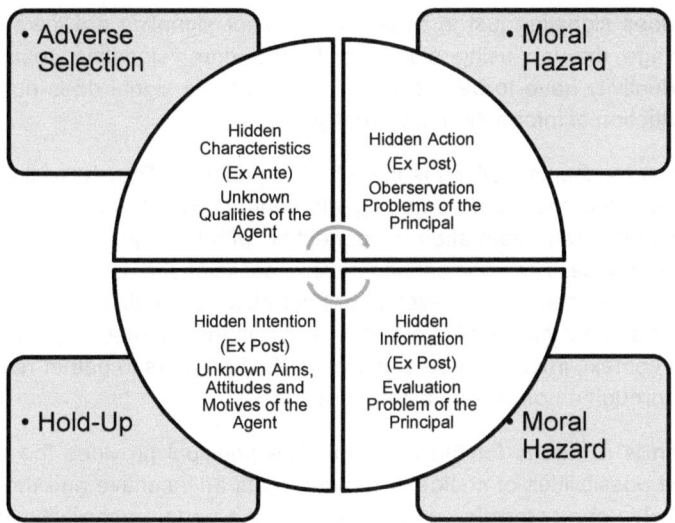

Figure 4: Overview of Information Asymmetry and Agency Problems

Own graphic according to: Erlei, M. et al. (2007); Laffont, J., Martimort, D. (2002), p. 3; Picot, A. et al. (2008), pp.74-75.

The principal-agent theory offers three different mechanisms to solve the problem of adverse selection. All mechanisms are aimed at reaching information equilibrium between principal and agent:[41]

1. Signaling,

2. Screening,

3. Self-selection.

Signaling is a mechanism which can be used by the agent to demonstrate their high capabilities for the task. In this way, the agent can differentiate from other

---

[40] Cp. Erlei, M. et al. (2007), pp. 69 and 103; Laffont, J., Martimort, D. (2002), p. 29; Picot, A. et al. (2008), pp. 76 and 142.

[41] Cp. Picot, A. et al. (2008), p. 76.

agents. There are different purposes for signaling. Due to the financial market, signaling appears often in context of credit negotiations. Classically, borrowers signal their unobservable default risk type by presenting several evidences for improper repayment of the loan. Means of signaling include contracts of previous loans, references from previous lenders, guarantees, etc.[42]

The agent does signaling just in case the costs for signaling are lower than their advantage through using signaling. Furthermore, signaling costs and agent's productivity have to be correlated. Otherwise, a signal does not contribute to reduction of information asymmetry.[43]

Screening is a mechanism which is used by the principal. They have the possibility to screen the agent before getting into contract with the agent. Ex ante, the principal gets more information in connection with the agent's characteristics and performance. The principal has several occasions to screen the agent. He can screen the agent in its presence or their absence by different tests, assessments or another media of information such as the internet. In general, in the financial context, investors use different media and ways to gather relevant information through a possible investment option.[44]

Self-selection is a means for the principal. The principal provides the agent with different possibilities of choice. The agent gets an incentive and the principal reveals his characteristics when they select a certain possibility. In the financial sector, self-selection is often used to get an overview of the agent's risk profile. Due to the option the agent takes, the principal gets a closer look at the general risk profile of the agent.[45]

Assimilation of interest is a further means to reduce information asymmetry, especially in connection with adverse selection. The aim of assimilation of interest is to approach the agent's and principal's interests. The agent is obliged to develop self-interest in the kind of business which they do. Through this development, the agent gives the principal ideas for needing services. Thus, the principal motivates the agent to provide good quality in work.[46]

There are two mechanisms to reduce the problem of moral hazard. On the one hand, there is the described assimilation of interest and the other is mentoring. When this behavior is connected to the subject of hidden information, princi-

---

[42]   Cp. Ben-Shahar, D., Feldman, D. (2003), pp. 173-174; Picot, A. et al. (2008), p. 78.
[43]   Cp. Spence, M. (1973), pp. 358-359.
[44]   Cp. Picot, A. et al. (2008), p. 78.
[45]   Cp. Picot, A. et al. (2008), p. 78.
[46]   Cp. Picot, A. et al. (2008), p. 78.

pal's incentives can motivate the agent. Incentives are, for example, participation in the outcome, any other kind of direct or indirect outcome-orientated financial rewards (i.e. bonus system, stock options, or additional vacation, etc.). In this connection, it is important that companies install a functional incentive and sanction system in company. Using such a system, the aspect of risk allocation between principal and agent can be considered.[47]

A lot of incentive systems are based on outcomes. Therefore, it is mutually exclusive from exogenous factors. Otherwise, there is a risk transfer from principal to agent. If the outcome is unsatisfactory, the principal will not be able to evaluate the agent's task. It can happen that the agent gets punished although they do their work well. This situation is very important in the context of expected salary. Especially, the members of the management board often receive outcome-based extra earnings. To motivate these people, it is necessary that their work is evaluated, not exogenous factors. Failing which, they do not work any longer in the principal's interest.[48]

Besides monetary incentives, there are several non-monetary incentives. One of these non-monetary incentives is related to corporate social responsibility. Implementing social aspects in company motivates agents. Hence, they have to be convinced that corporate social responsibility is important for society. If it is true, agents will be motivated when company adopts environmentally friendly technologies. Corporate social responsibility can be positive for both agent's and principal's motivation.[49]

Monitoring systems supports the principal detecting the agent's hidden actions. Systems can focus on formal planning, budgeting, accounting systems, reporting procedures, additional layers of management and control, as well as risk management systems. The amount of information the principal is able to gather through monitoring has influence on the agent's behavior. A fortiori, the agent behaves in the principal's interest.[50]

Assimilation of interest is also an approach to solving the problem of hold-up. Therefore, the principal needs securities like guarantees, countertrade, or in the best way the agent's reputation. By using a security, the agent is inhibited to act against the principal's interest. Besides, the one-way dependency of the

---

[47] Cp. Eisenhardt, K. (1989), pp. 60-61; Erlei, M. et al. (2007), p. 71; Picot, A. et al. (2008), p. 79.
[48] Cp. Besley, T., Ghatak, M. (2005), p. 161; Eisenhardt, K. (1989), pp. 60-61.
[49] Cp. Besley, T., Ghatak, M. (2005), pp. 158-161.
[50] Cp. Eisenhardt, K. (1989), pp. 60-61; Picot, A. et al. (2008), p. 79.

principal is on a lower level. The agent has their own interest in a successful business.[51]

Institutional integration is another possible solution to hidden intention. Its origin is in the transaction cost theory. Suppliers and customers are very important for a company. In particular, suppliers are relevant for supplying important goods. To lower opportunistic behavior on the vertical aces, long-term binding contracts, equity participation, and other forms of vertical integration result in tighter control by the principal and less leeway for the agent.[52]

Principal-agent research helps to look more closely at the importance of information as commodity, incentives, self-interest, and risk allocation in organizational thinking. Hence, it is important to analyze the agent's as well as principal's role in organizational thinking. It supports defining a functional basis of information systems within a company. Moreover, outcome-oriented contracts help to control opportunism. So, it is possible to design an institution which functions in the most efficient way on a given situation.[53]

The following overview is a summary of the problems and recommendations of the principal-agent theory due to financial relationships. Furthermore, it focuses on the methods and means of resolution for one agency problem:

Agency Problem	Adverse Selection	Moral Hazard		Hold-Up
Agent's Leeway	Ex ante	Ex post		Ex post
Information Asymmetry	Hidden characteristics	Hidden information	Hidden action	Hidden intention
Principal's Problem	Uncertainty about agent's qualities	Uncertainty about agent's contribution to outcome		Uncertainty about agent's motives
Agent's Behavior	Fraudulent concealment and malicious deceit	Deliberate reluctance		Planning of principal's dependency

---

[51] Cp. Picot, A. et al. (2008), p. 80.
[52] Cp. Grossmann, S., Hart, O. (1986), p. 716; Picot, A. et al. (2008), p. 80 and 124.
[53] Cp. Eisenhardt, K. (1989), pp. 64-69.

Agency Problem	Adverse Selection		Moral Hazard		Hold-Up
Example	Creditor and debtor		Risky business operations		Specificity of an investment
Method of resolution	Elimination of information asymmetry through signaling, screening and self-selection	Assimilation of interest	Assimilation of interest	Reduction of information asymmetry	Assimilation of interest
Means of resolution	Certificates, contracts of previous loans, references from previous creditors or guarantees	Agent's reputation	Participation on the outcome, any other kind of direct or indirect outcome-orientated financial reward (i.e. bonus system, stock options or additional vacation)	Monitoring, reporting and control systems	Long-term binding contracts, equity participation or another form of vertical integration

Table 1: Problems and Recommendations of Principal-Agent Theory

Own table according to: Picot, A. et al. (2008), p. 77.

3 Financial Communication in Small and Medium-Sized Enterprises

Regarding the main problem, Chapter 3 gives an overview of the special characteristics of German small and medium-sized enterprises. Therefore, the general financing conditions of these companies are presented. It is followed by requirements which are important due to the changes brought about by the financial crisis.

3.1 Characteristics of German Small and Mid-Caps

SMEs are often unknown in public but they are very important for economy in general. Compared to the whole economy, SMEs are the biggest employers. SMEs offer jobs in all business areas. For example, they offer 60% of all jobs within the manufacturing sector.[54]

German SMEs do their bit for Germany's economic growth. In comparison with the United States and the United Kingdom, the SMEs in Germany are more important for the whole economy than the big companies. In Germany, more than 99% of all companies are small and medium-sized enterprises.[55] This is the reason for the importance of these companies for Germany's economy.

SMEs are the innovative suppliers of the bigger companies and offer most of the jobs. Additionally, they support economic stability because the owners of the SMEs are often the managers of the companies at the same time. So, their personal insurances are covered by good development within their companies. German SMEs are an example of good value-driven management.

In comparison with big companies, there are differences in quantitative and qualitative criteria. In particular, the organizational structure differs from that in big companies.[56] These differences have consequence for all parts of an SME. The culture as well as the processes is different in these companies.

The number of employees, the sales, and the balance sheet total are factors belonging to the quantitative criteria to mark out companies which are small

---

[54] Cp. Ayyagari, M. et al. (2007), pp. 415-434.
[55] Cp. Audretsch, D., Elston, J. (1997), p. 97.
[56] Cp. Krämer, W. (2003), pp. 8 ff.

and medium-sized. Quantitative criteria are very important in connection with subventions and other benefits for companies. For most sponsorship, the EU definition of small and medium-sized companies is relevant:

EU Definition	Micro	Small	Middle	Big
Employees	≤ 9	≤ 49	≤ 249	≥ 250
Sales	≤ 2 Mio. EUR	≤ 10 Mio. EUR	≤ 50 Mio. EUR	≥ 50 Mio. EUR
Balance Sheet Total	≤ 2 Mio. EUR	≤ 10 Mio. EUR	≤ 43 Mio. EUR	≥ 43 Mio. EUR

Table 2: Quantitative Criteria of SMEs

Own table according to: European Commission (2005), p. 14.

Additionally, there are several qualitative criteria for a company to verify whether the company is an SME. In this context, the following criteria are important:[57]

- Company's manager is often the owner of the company,
- Company's owner shapes the culture in the company,
- Company's management is part of a functional network and has good connections with customers and suppliers,
- The hierarchy in the company is flat,
- Management and employees work closely together,
- The choice of products is not so diverse as in big companies,
- Company's structure can be fast fitted due to changes in the company's environment,
- Financial and personal resources are bonded,

---

[57] Cp. Flueglistaller, U. (2004), p. 11; Gelbmann, U. et al. (2004), pp. 252-255; Herstatt, C. et al. (2001), p. 150; Minder, S. (2001), pp. 8-11; Mugler, J. (2005), p. 17; Pfohl, H. (2006), p. 18-21; Walther, S. (2004), pp. 36-37.

- Wrong decisions can threaten the company's survival,

- Planning and controlling are underdeveloped,

- Decisions are often based on intuitions.

Small and medium-sized enterprises do not fulfill all criteria. It is important that they fulfill some of these criteria to belong to the SME category.[58]

The characteristics of SMEs can be an advantage or a disadvantage for the company and its employees. Firstly, there can be a mentionable disadvantage. SMEs offer most jobs within an economy, but at the same time destroy most jobs. The fixed structure of SMEs often leads to the point where they are unable to survive in the changing environment. SMEs do not have plenty of staff to consider all legal and economic changes. Therefore, SMEs often have to adopt developments which are initiated by bigger companies.[59]

According to the mentioned characteristics of SMEs, the access to financial means is often limited. They have to offer more securities and pay more interests on debt capital, because the probability of default is higher than in bigger companies. These factors reduce the power of an SME in the way of creating new values. A stable financial background and the possibility to finance from its own earnings are very important for an SME because they often get no external capital.[60] Further financial related aspects will be discussed in the following chapter.

Otherwise the nonrigid structures offer the chance creating innovative and different ideas. SMEs are very important concerning innovations and technologies.[61] SMEs have to be creative and innovative according to the company's development. Development relies on the grade of innovation because that is the only possibility to survive in the competition. SMEs have to be innovative to get interesting for big companies as business partners and finally as supplier.

The probability of insolvency is often related to the development of the geographical region where the SME is located. If there is good economic, social, and political development in one reason, SMEs located in this area also devel-

---

[58] Cp. Mugler, J. (2005), p. 17.
[59] Cp. Davis, S. et al. (1996), w/o p.
[60] Cp. Beck, T. et al. (2005), pp. 137-177; Beck, T. et al. (2006), pp. 932-952; IADB (2004), w/o p.
[61] Cp. Acs, Z., Audretsch, D. (1990), w/o p.

op in a positive way.[62] SMEs often interact closer than bigger firms with other companies which are nearly located to the firm. Furthermore, they profit directly from investments in infrastructure and other things which improve the location. They need external support because in comparison with big companies, they are unable to build all structures on their own. This also includes offerings for the leisure time as well as for childcare.

In later empirical analysis, just a small part of all SMEs can be analyzed. Owing to methodological reasons, the empirical analysis focuses on SMEs which are listed on a German stock exchange. Furthermore, according to comparable reasons, they have to belong to the prime standard of the German stock exchange.

3.2 Financing in SMEs

Small and medium-sized enterprises differ from big companies in several points. However, some points are similar to big companies. Due to financial issues and the ability to pay redemption and interests on liabilities, SMEs have to signal in a more powerful and sustainable way than big companies that they are able to pay redemption and interests.[63]

Furthermore, SME financing has changed since the start of the financial crisis. SME financing is in an ongoing process of change in the part of regulation in consequence of the financial crisis. The crisis induces more regulatory requirements for banks in the field of SME financing. In particular, the lending process of banks is touched by these developments.[64]

Today, the credit volume is lower than what it was before the financial crisis. Before the crisis, the regulatory requirements for lending capital were lower, which could be a reason for the decrease. Currently, lenders have to fulfill a lot of requirements in the process of granting a loan. These have consequences for SMEs as well as for banks. SMEs have to look closely at financial issues to ensure a constant refinancing for the operating business. Banks have to concentrate on lending processes within their organizations. Old processes have to be analyzed and modified because of the new requirements.[65]

[62] Cp. Lehmann, E. et al. (2004), p. 28.
[63] Cp. Pindado, J., Rodrigues, L. (2004), pp. 51-66.
[64] Cp. Gambacorta, L., Marques-Ibanez, D. (2011), pp. 140-147.
[65] Cp. Wehinger, G. (2012), p. 66.

The new requirements are connected to capital and liquidity issues. A general deleveraging is the aim. The new requirements are imposed under Basel III which is a further step than Basel I and II.[66] Basel III primarily includes three changes in contrast to Basel II. Firstly, the quantitative and qualitative requirements on capital are higher. The equity capital of a bank is based on three different classes of capital—hard capital (Core Tier 1), soft capital (Tier 1), and supplementary capital (Tier 2). Furthermore, there are new requirements about liquidity and leverage ratio.[67]

Essentially, SMEs have to deal with the first mentioned point as banks have to withhold more equity when they grant credit to a company with a bad rating to save their own. This leads banks to choose companies in credit business with a good rating. As a result, companies with bad ratings have to search for other refinancing possibilities. Traditionally, SMEs have a worse rating in comparison with big companies due to their organizational structure.[68]

Today, there are other important actors besides banks. On one side, these actors are competitors for the banks; on the other, they complete banks' offers. The actors support to fulfill the credit needs of the economy. In addition to traditional bank financing, market financing becomes more and more important.[69]

In comparison with the United States and the United Kingdom, SMEs are strongly dependent on banks for external finance. The capital market is relatively undeveloped in Germany. In the past, Germans had a close relationship with their house banks. The house bank was a partner for all financial issues. German entrepreneurs do not like to discuss financial issues. Especially, they do not want to talk about these things in public. That is one reason why German SMEs do not prefer joining the capital market for refinancing. However, in the aftermath of the financial crisis, SMEs have to look to the capital market for refinancing.[70]

The special German financial system, comprising different bank types, has special importance for refinancing of SMEs. In Germany, there are a lot of small and medium-sized savings banks and cooperative banks. These banks are mostly the only ones for all financial issues of the SMEs. Just less than one-tenth of all German banking assets are accounted for by the bigger private

---

[66] Cp. Wehinger, G. (2012), pp. 68-69.
[67] Cp. BIS (2010), w/o p.
[68] Cp. Gambacorta, L., Marques-Ibanez, D. (2011), pp. 140-147; Pindado, J., Rodrigues, L. (2004), pp. 51-66.
[69] Cp. Wehinger, G. (2012), p. 66.
[70] Cp. Audretsch, D., Elston, J. (1997), p. 97.

banks. So, the financial partner is directly from the immediate environment of the company. This improves the relationship between bank and customer. Simultaneously, it can be bad for the relationship in bad times because the bank has more relevant information which can threaten the company.[71]

The close relationship between SME and house bank leads to strong dependency of the SME on the house bank. The house bank is able to design contracts at its discretion.[72] Because of this, banks can determine collateral requirements and loan prices without paying too much attention to companies' ideas based on the loan. Bank's self-interest outweighs company's needs.

Theoretically, banks have to analyze all information from the relationship to the SME to resolve credit rationing problems. In reality, banks use information asymmetry to gather more than the relevant information about the SME. Based on the wide range of information, banks are able to measure SMEs' default risks by internal risk classes or borrowing ratings. Rating practices differ from bank to bank.[73] They are not transparent for clients like SMEs. Clients are prone to give banks more information during the refinancing process than necessary, leading to disadvantages on the client's side because banks use the information for justifying higher interests.

With regarding to the principal-agent approach, there are also information asymmetries the other way round. In comparison, bigger companies are more transparent than SMEs. Furthermore, in big companies, there is a clear separation between owner and management. In SMEs, the owner is often the manager of the company. The owner tries to optimize all things related to financial issues in their own way because their existence depends on the company. Therefore, the owner tries to hide all information that can threaten the refinancing of the company. Due to the mentioned agency problem of moral hazard, bank and other investors try to eliminate the information asymmetry through monitoring. Because of lower level of available information, SMEs have to signal more than big companies their capabilities for payment of redemption and interests.[74]

On the one hand, banks profit from intensive internal rating practices. There are decreasing risks due to insolvency and falling out redemptions of SMEs. On the other hand, SMEs have to pay more interests on loans than other

---

[71]   Cp. Audretsch, D., Elston, J. (1997), p. 102.
[72]   Cp. Carlin, W., Richthofen, P. (1995), p. 20.
[73]   Cp. Basel Committee on Banking Supervision (2000), w/o p.
[74]   Cp. Bartscherer, M. (2004), pp. 44-69; Huchzermeier, M. (2006), pp. 22-24.

companies which are independent of one house bank. This leads to another important disadvantage for SMEs in the context of refinancing. Increasing loan rates are connected with decreasing collaterals. SMEs have to provide more collateral to decrease their interest expenses than bigger companies with a better rating and cheaper conditions for refinancing.[75]

For refinancing, too intensive relationships between banks and SMEs are bad. Particularly, in this case, the SME has to finance a risky project for a short period. Here, the bank notes all bad things from the relationship which additionally increases the interest rate. In general, a close relationship between banks and SMEs multiplies in times of bad financial situation; SME's interest expenses threaten the whole business of the SME.[76]

In this context, the assessment of creditworthiness of SMEs is complex because of the close relationship between banks and SMEs in relation to other loan types such as mortgage loans. Today's SMEs are active in an international business environment, but simultaneously they also have a bounded and often specialized product portfolio. This complexity in SME's business has as consequence that banks have problems to evaluate SME's profitability of failure. In addition, it is difficult for banks to find adequate securitization for a loan. The specification within an SME is a cost driver in the process of giving a loan because banks have to do a detailed cost-intensive due diligence. The costs for the due diligence have to be refinanced through the accommodation of a loan, but this is often difficult because the loan size in SMEs is relatively small. For dealing with this problem, SMEs have to look for other possibilities of refinancing. In the past, mezzanine capital was an often used means.[77]

The specifications in SME financing need adjustments within the financial system and within one banking group. These needs had influences on the structure of German banks, especially on savings and cooperative banks. These banks are the most important lenders for SMEs. There are different institutions which are implemented in the financial system to contain banks' risks and to grant SMEs better credit conditions:[78]

1. There are special credit institutions such as the KfW which are implemented by government to give SMEs the chance to get long-term fixed-rate loans on good credit conditions,

---

[75]   Cp. Lehmann, E. et al. (2004), p. 25.
[76]   Cp. Lehmann, E. et al. (2004), p. 25.
[77]   Cp. Wehinger, G. (2012), p. 75.
[78]   Cp. Audretsch, D., Elston, J. (1997), p. 104.

2. Special subsidy programs to strengthen the financial background of SMEs for further investments,

3. Special refinancing and risk pooling mechanisms offered by banks and insurances to share risks.

Two of three long-term loans to SMEs have one of these three possibilities.[79]

These means of supporting SMEs were implemented after the Second World War to facilitate reconstruction in Germany. During this time, German SMEs should develop technological competence to allow competitiveness in the international business environment.[80]

Moreover, mezzanine capital and private equity are important in SME financing. These financial means are able to take risks out of bank balance sheets. Mezzanine capital and private equity investors are prepared to take more risks than banks.[81] These investors look for high yields, which is why they accept higher risks. They evaluate risks in another way because they do not just focus on securitization. They also look at the business model and possible future earnings.

Moreover, there are structured financial products in the capital market especially for SMEs. SMEs often issue covered bonds to actors from the capital market. These bonds are covered with mortgages and other securitizations. In contrast to mezzanine capital, bonds do not strengthen an SME's equity base, but a successful placement in the capital market improves the standing of the SME. Bonds are practical means for banks to shrink risks in their balance sheets. The effect on balance sheet is positive for banks because they can earn money based on the placement and simultaneously save money because they do not have to hold up equity capital owing to Basel III. But covered bonds are debt capital for companies and this fact is bad in comparison with mezzanine capital. Companies cannot decrease the debt equity ratio which has negative impacts on further bank financing.[82]

On the basis of the theoretical background, SMEs have to create a trustworthy relationship to their investors. Therefore, it is irrelevant whether the investor is a bank or somebody else. In the following chapter, the possibilities of creating such a relationship are shown.

---

[79]   Cp. Audretsch, D., Elston, J. (1997), p. 104.
[80]   Cp. Audretsch, D., Elston, J. (1997), p. 104.
[81]   Cp. Wehinger, G. (2012), p. 67, 73.
[82]   Cp. Wehinger, G. (2012), p. 74.

3.3   Financial Communication Partners for SMEs

SMEs have to change their communication practice to meet investors' demand. On the one site the financial crisis and the lack of trust is a reason for
the changed demand. On the other site there are other actors which are important for SME's refinancing, e.g. participants on capital markets or investors
from abroad. Traditionally, banks were the most important partner for external
finance in the SME sector. Today, they are just one of a lot of other players
who are interesting for SME's refinancing possibility.[83]

At the moment banks are still the lender with the highest amount of SME financing. The European economy is about 75-80% bank financed. In the USA,
the situation is different because there are the capital market has a better
standing and this influences its importance due to SME financing. There, the
economy is only about 25% bank financed. The rest is financed by the capital
market.[84]

Another distinction which is based on the point before is the close relationship
between banks and SMEs in Germany. In Germany, bankers and entrepreneurs maintain an intensive partnership. That is one reason for the good positioning of banks in German SMEs. In the past, bankers were extensively represented on SMEs' supervisory boards. This strong connection reinforces the
foundation of trust. But otherwise it pushes a kind of internal capital market
which makes the relationship for externals opaque. The opaqueness leads to
the point that there is often one close financial relationship between one SME
and one bank. For other financial partners the relationship is not transparent
and too risky. These partners do not focus any longer a relationship.[85]

This established practice leads to the fact that SMEs with a long history and
close relationship between SME and bank for a long time have a better position at their house banks than young companies. This has different consequences for the SME sector in Germany. On the one site companies with an
intensive relationship and a constantly communication receive better credit
conditions because the close relationship strengthen the foundation of trust.
Young companies have to build up such a close relationship. Reaching this

---

[83]   Cp. Gambacorta, L., Marques-Ibanez, D. (2011), pp. 142-149.
[84]   Cp. Wehinger, G. (2012), p. 67.
[85]   Cp. Cable, J. (1985), p. 119.

relationship it is important that young companies improve their communication practice concerning the needs of the bank.[86]

Particularly for the German SME sector, there was a good example of the mentioned points in the recent past. After reunification of East and West Germany, eastern firms had their difficulties due to refinancing. There was no foundation of trust between banks and SME. The relationship was not close enough to get the same credit conditions like western firms. This leads to the point that eastern companies had to offer more collateral to get adequate and with western firms' comparable credit conditions.[87] Financial communication between SME and bank improve the relationship and offers SME the possibility to better conditions. Furthermore, the communication has to be on a high level over years. In case of eastern and western German companies, they were on the same level after a period of six years due to the collateral[88].

Due to the theoretical background of the existing information asymmetries, the owner of a company and the management do not have the same interests. Both have the aim to maximize their own benefits. Making a functional contract between the opponents causes costs. Especially in the field of SMEs, these costs are higher than in big companies because the monitoring of SME is more difficult concerning the in comparison to big companies opaquely structure of SMEs.[89]

Young and small firms have to communicate more intensive with investors than established and bigger companies. The young and small enterprises learn through new projects how to deal with capital and they realize the sustainability in their businesses in practice. At the beginning these companies have high growth rates and fluctuating cash flows. Therefore, these companies have to pay higher credit conditions when they do not influence their conditions through purposeful communication with the investors. Concerning this, how to communicate with investors also depends from firm's characteristics. Today's investors are very interested in cash flows. In the past, the earnings were the weighty figure in a corporation's evaluation. In contrast to cash flows, earnings are more easily influenced by company. Consequently, young and

---

[86]   Cp. Becchetti, L. et al. (2010), p. 483.
[87]   Cp. Wehinger, G. (2012), p. 67.
[88]   Cp. Wehinger, G. (2012), p. 67.
[89]   Cp. Gerke, W. (2005), pp. 263-264; Grüning, M. (2011), p. 181; Hubig, C., Siemoneit, O. (2009), pp. 68-69.

small firms have to argue conclusively their historical and expected cash flows within their equity or debt story.[90]

The equity or debt story is an answer to the theoretical belongings of financial communication. The financial communication has to be objective, trustworthy, future-orientated, open and completed to generate a trustful relationship between management and investor. Due to this theory, SMEs have to select their investors and they have to communicate with them to ensure company's the further refinancing. It is equal whether the investor is a bank or somebody else.[91]

In this context, it is important to focus communication on the target group. All stakeholders have their special communication needs. These needs can be based on hard and on soft information. Therefore, SMEs have to get into dialogue to all communication partners for getting to know which information they need. In the context of finance, it is important that SME differentiate between each investor. All investors have different information needs.[92]

A good financial communication improves SME's and investor's level of information. The level of information is influenced by continuous, personalized and direct communication between SME and bank. These points do not affect the financial communication alone. They are important for all stakeholders.[93]

For the long-term success in part of refinancing, besides the content of information SMEs have to control the financial network. SMEs have to improve their financial network to guarantee good refinancing possibilities for the future. A tight relationship to the local house bank for example improves how mentioned before the refinancing options at the house bank but they are not good for the financial relations to all investors in general. SMEs with large capital demand profit from a wide-ranging financial communication at an early stage. A non-wide-ranging financial communication in the run-up to the first placement on the capital markets demonstrates firm's weakness. The bounded communication put other investors off because they do not build up the neces-

---

[90] Cp. Alti, A. (2003), pp. 707-722.
[91] Cp. Hillmann, M. (2011), p. 53; Hubig, C., Siemoneit, O. (2009), p. 65; Picot, A., Reichwald, R., Wigand, R. (2003), p. 123; Piwinger, M. (2009), p. 18.
[92] Cp. Pedersen, A. (2008), pp. 371-372.
[93] Cp. Berger, A. et al. (1995), pp. 155-219; Berger, A. et al. (2001), pp. 2127-2167; Keeton, W. (1995), pp. 45-57; Berger, A., Udell, G. (1996), pp. 559-627; Strahan, P., Weston, J. (1996), pp. 1-6; Mian, A. (2006), pp. 1465-1505; Sengupta, R. (2007), pp. 502-528.

sary foundation of trust to invest in the SME.[94] Therefore, SMEs have to plan their financial needs with a long-term focus. Due to this point, it is necessary that SMEs with a huge demand on capital in future focus the capital markets some time before they will join the markets. Early financial communication improves the foundation of trust between SME and investor in the long run and the SME has better chances to find investors in case of further placements.

More importantly, SMEs have to concentrate within their financial communication to all lenders. With a wide-ranging financial communication companies reach a higher number of lenders. These lead to better credit conditions because there is a positive effect between the number of lenders and the credit condition. Due to this point, companies which concentrate their financial communication to a huge number of investors train their communication skills. All investors need different information and companies with a wide-ranging communication learn to lower the relative production of information in equilibrium. So, it is important for a company to focus on financial communication at early stage in company's life cycle to reduce capital costs sustainably.[95]

Financial communication includes all mentioned theoretic means to minimize the information asymmetries between management and investor. The management has to signal the investor that he is the right person manage his capital. Therefore, it is important that the management gives all relevant information to the investor. Examples are all financial reports, information due to company's strategy and descriptions to the current situation of the company.[96] Orientation offers the requirements of market standards with a higher level of transparency. One example is the German prime standard. Companies from the prime standard have to offer investors an online calendar with all relevant dates for the company, possibility for joining a conference and they have to inform investors about relevant changes through Ad-hoc announcements.[97]

The changes in SME financing from bank lending to capital lending need changes in several things, also in practice of financial communication. The change is easier in the United States than in Europe because there are 75% financed by capital markets. In Europe, the situation is different and it takes

---

[94]   Cp. Becchetti, L. et al. (2010), p. 481.
[95]   Cp. Becchetti, L. et al. (2010), p. 483; Bolton, P., Scharfstein, D. (1996), pp. 1-25; Detragiache, E. et al. (2000), pp. 1133-1161; Petersen, M., Rajan, R. (1994), pp. 1367-1400; Thadden, E. (1995), pp. 557-575.
[96]   Cp. Grüning, M. (2011), p. 59.
[97]   Cp. Deutsche Börse (2012), pp. 1-4.

more time until firms take the capital market as an equal complement to classical bank lending.[98]

In general, the changes in refinancing practice depends further changes in all parts. Also the regulatory part has to be changed. There, the industry and the regulators have to find new securitizations which concentrate on the opacity and complexity of new financial products.[99] Companies can assistant the process through open and trustworthy financial communication which addresses all needs of the investors.

Financial communication's main aim is to handle the information asymmetries between company owner and investor. Therefore, it is necessary that the financial communication increases the trust in management's competences, reaching profit and return targets, and trust in management's intention, omission of opportunistic behavior.[100]

Summarizing, the financial communication is important for SME's long-term success. Due to the success, it is important that companies discover the contents within the financial communication which are important for their investors. The contents can differentiate on base of several things like business are, company size, number of employees and so on.

---

[98]   Cp. Wehinger, G. (2012), p. 71.
[99]   Cp. Wehinger, G. (2012), p. 74.
[100] Cp. Hubig, C., Siemoneit, O. (2009), pp. 68-71.

4 Patents in Financial Communication

Chapter four is directly linked to the main problem. Firstly, the different types of patents are presented. After it, the German patent application process is described. In the last part of this chapter, patents as an example of intangible assets are presented.

4.1 Types of Patents

Patents are based on inventions. An invention can take place in technical and other parts. It is the realization of a new problem solving. The characteristics of the problem solving approach can be different. The discovering of an invention can be determined or not. Often, the solution is delivered through practical experiences.[101]

The innovation typically follows the invention. In contrast to the invention, the innovation is the first economic and aware use of the invention in business. There are a lot of inventions in firms which are not important for company's business. In these cases, the invention remains an invention because for the company it is unattractive to public the invention as an innovation. There is no economic use for the company.[102]

Additionally to invention and innovation, there is the imitation. Imitation follows normally an innovation. If a company presents new products or problem solving processes, other companies imitate these products because otherwise they lose their customers. Without imitation these firms are less attractive for customers than other competitors.[103]

Imitations are not negative in general. They are normally little varieties of the original product or problem solving approach. These little differences lead to new ideas according to the product or the problem solving process. That pushes the development of all products and problem solving approaches. All

---

[101]   Cp. Perl, E. (2007), p. 20-23.
[102]   Cp. Specht, G. et al. (2002), p. 13.
[103]   Cp. Vahs, D., Burmester, R. (2005), p. 80.

companies can learn from each other when they get to know details competitor's products and problem solving processes.[104]

The described possibility to generate a benefit for all engaged companies, is the reason for a lot of cooperation in field of Research and Development (R&D). Cooperating firms have the chance to profit from knowledge spillovers. Furthermore, there is less uncertainty due to costs and processes. Cooperating firms have the probability to share costs for R&D. Besides the costs, they can also share their experiences in R&D.[105]

Patents are important in all industries. There are some industries which benefit disproportionately from patents. Companies within these industries have more innovative opportunities and patents are more important for these companies for positioning themselves in competition.[106]

In Germany, the legal protection of intellectual property rights belongs to the legal protection of industrial property rights. Simultaneously, it belongs to the civil law. The aim of the legal protection is to protect intellectual work. The industrial property rights are based on several sources of law from different fields of law: Patent law, utility model right, semiconductor property right, taste pattern right, unfair competition law, trademark law.[107]

The later following empirical analysis is based on data from European Patent Office. The European and the German patent system hardly differ from one another. European patent system's origin is the German patent system. Therefore, both systems can be considered simultaneously.[108]

Patenting is based on technological or procedural new inventions. Furthermore, for patenting it is necessary that the invention offers the possibility for a commercial utilization. On this occasion just the application of the invention is relevant. Whether the invention will be realized or profitable in future is not relevant.[109] An invention is than new when it is better than the current state of the

---

[104] Cp. Albach, H. (1990), p. 97; Bähr-Seppelfricke, U. (1999), w/o p; Rogers, E. (2003), w/o p.

[105] Cp. Combs, K. (1992), pp. 353-371; d'Aspremont, C., Jacquemin, A (1988), pp. 1133-1137; Kamien, M. et al.. (1992), pp. 1293-1306; Marjit, S. (1991), pp. 187-191; Mukherjee, A., Marjit, S. (2004), pp. 243-258; Suzumura, K. (1992), pp. 1307-1320.

[106] Cp. Cohen, W. (1995), w/ p; Cohen, W., Levin, R. (1989), w/o p; Pakes, A., Schankerman, M. (1984), w/o p.

[107] Cp. Ahlert, D., Schröder, H. (1996), p. 105; Osterrieth, C. (2004), p. 89.

[108] Cp. Reitzig, M. (2002), p. 8.

[109] Cp. Ensthaler, J., Strübbe, K. (2006), pp. 12-13; EPÜ (2000), art. 52, para. 1; PatG (2011), § 1 para. 1.

art. In this case, the state of the art covers all technological and procedural innovations which are written, oral or through practical use known to the public until the application date. Additionally, it is important that the innovation is absolutely new. The innovation has to be enrichment. The examination for novelty happens regarding to the average knowledge of an expert. Relevant for patenting is the technological or the procedural state of the art at the application date.[110]

Patenting is possible for technological and procedural inventions. Generally, there is a distinction between product patents and process patents. Product patents offer a wide-ranging protection because they cover all steps of the manufacturing process as well as all possibilities of use due to the invention. Product patents are granted for the following things:[111]

1. Machines, devices, tools other than just parts of the mentioned ones,

2. Not natural substances like plastics, pharmaceutical products and so on,

3. Immobile things like bridges, dikes or canals,

4. Electrical circuits and other functional interacting things.

Process patents include use claims. They are granted for invention due to the following points:[112]

1. Manufacturing methods for the production,

2. Working methods for several practical uses which do not belong to a special production,

3. New possibilities to use a product in an unknown way.

Since 1975 the Strasbourg Agreement Concerning the International Patent Classification is the most important orientation for patent research. It classifies the patents to eight different classes and a lot of subclasses. For the further research, just the main subdivisions are relevant. These subdivisions are:[113]

A. Human Necessities

---

[110] Cp. EPÜ (2000), art. 54, para. 2; PatG (2011), § 3 para. 2; Osterrieth, C. (2004), pp. 104-110.
[111] Cp. Ensthaler, J., Strübbe, K. (2006), p. 13; Pleschak, F., Sabisch, H. (1996), p. 49.
[112] Cp. Ensthaler, J., Strübbe, K. (2006), p. 13
[113] Cp. Bittelmeyer, C. (2007), p. 120.

B. Performing Operations, Transporting

C. Chemistry, Metallurgy

D. Textiles, Paper

E. Fixed Constructions

F. Mechanical Engineering, Lighting, Heating, Weapons

G. Physics

H. Electricity

4.2   Process of Patent Application

The starting point of the process of patent application is hard to determine. Firstly, patents are the written document of an innovation which was invented in firm. Innovations can be analyzed in a procedural or object orientated view. The procedural view focuses on the process of generating an innovation. The object orientated view concentrates on the results of the innovation.[114] Patents are in this context the finished result of an innovation process. Therefore, it is important to differentiate between invention, innovation and imitation. All these points belong to the general patent topic.

Typically, innovations are separated in product innovations, process innovations and social innovations. Product innovations are material or immaterial things which can be sold on markets to other market participants. Process innovations appear in one company or within a supply chain. They define process in a new and more efficient way in relation to the old processes.[115]

The revelation of the invention is necessary requirement to get a protection against improper commercial and general use of the invention. Without revelation patenting is impossible. The protection is only possible when the inventor agrees to reveal all details of his invention. More precise, the inventor has to present these details of his invention which another expert needs to use the invention. But, it is also important that the invention is present to the general public. The invention is a development of the current state of the art and a

---

[114]   Cp. Ahsen, A. et al. pp. 5-6; Macharzina, K., Wolf, J. (2008), p. 726.
[115]   Cp. Hauschildt, J., Salomo, S. (2007), p. 9.

general progress for the society. Everybody needs access to the invention to improve society's development.[116]

The process of patent application and publication of the patent need a lot of time. From application to publication it often takes 18 months. That can be a problem because a plenty of information due to patents are old and do not represent the current state of the art of technological and procedural development. This point is critical for company's competitor but also for other stakeholders like investors. If these people take the patents under consideration, they will have a distorted result of enterprise's value. The amount of information and the value of intellectual property are unknown for a long period of time.[117]

The patent specification includes applicant's name, request for the grant of a patent, patent claim, patent description and all drawing due to the patent. The patent claim is as short as necessary. That is important to see the innovation on the first view. Lacks can be repaired later but it is impossible to add missing contents.[118]

The following graphic displays the typical way from patent application until patent publication. In the graphic, there are all possible steps within the whole process:

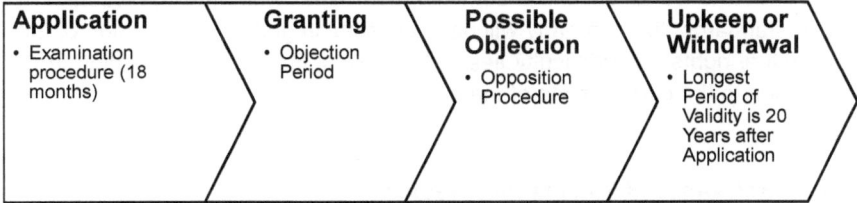

Application
• Examination procedure (18 months)

Granting
• Objection Period

Possible Objection
• Opposition Procedure

Upkeep or Withdrawal
• Longest Period of Validity is 20 Years after Application

Figure 5: Chronology of German Patent System
Own graphic according to: Cp. Ensthaler, J., Strübbe, K. (2006), p. 17

After patent application, the patent office checks formal criteria. Besides, it looks whether there are some obvious obstacles for patenting. Besides these checks, patent office assigns the invention to a technical field.[119]

---

[116] Cp. EPÜ (2000), art. 83; PatG (2011), § 34, para. 2; Osterrieth, C. (2004), p. 16.
[117] Cp. Gzybowski, M. (2013), p. 9.
[118] Cp. PatG (2011), §§ 34-41; Rebel, D. (2001), p. 342.
[119] Cp. Schmoch, U. (1990), p. 16.

If all criteria are square, the patent office will publish the invention after 18 months to the public. The long period of time in this context is based on fundamentals of the patent law. The patent law offers a chance to the inventors with the long period of time. The inventors have the chance to use their invention on their own in this time. They can test the invention in business and the competitors do not know from the invention.[120]

Competitors and others have the possibility to object a patent within three months after the patent publication. In that case, granting the patent is examined again. The longest possible term of protection through a patent is 20 years after the date of patent application. The patent holder has to pay a fee for the patent after the third year. If he does not pay the fee, the patent expires and there is no protection any longer.[121]

Subsuming, the patent system is Germany is similar to other European states. There are just some specifications within the national patent systems. These specifications are not important for the later analysis because they are marginal.[122] In practice, it is difficult for companies to get protection through a patent in several countries. There is no international patent which is guilty in all countries. Companies have to apply a patent in all countries where they need the protection.[123]

The process of patent application is not difficult. There are minimal standards which have to be taken under consideration. If a company pays attention to these standards, the company will be successful in getting a protection of their intellectual rights. The difficulty lies in the description of the innovation in demarcation to the current state of the art.[124]

4.3   Relevance of Patents for Investors

The expenditures for R&D increase proportionally to company's size. But, very small and large firms are the most R&D intensive companies. Very small companies want to position themselves in a good position in competition. Large firms want to hold their positions in competition. To reach their aims, the firms invest a lot of money in R&D to be innovative and attractive for all stakehold-

---

[120]   Cp. Ensthaler, J., Strübbe, K. (2006), p. 18.
[121]   Cp. Schmoch, U. (1990), p. 16.
[122]   Cp. Haugg, N. et al. (1989), pp. 183-189; Specht, G. et al. (2002), p. 247.
[123]   Cp. Ensthaler, J., Strübbe, K. (2006), pp. 19-20; Schmoch, U. (1990), pp. 23-25.
[124]   Cp. Griliches, Z. (1990), p. 1690.

ers. Investors belong to the stakeholders. In relation to company's peer group, company's R&D expenditures are an indicator for innovativeness of the company.[125]

Furthermore, the R&D expenditures are linked to the sales of a company. Companies with increasing sales have higher R&D expenditures. The R&D expenditures strengthen the company's innovativeness. This point leads to better sales and at the end earnings in company's future. That is another important factor for investors to evaluate company's future and their own possibilities of future profits with an investment in these companies.[126]

In connection to the principal-agent model, there is discussion whether internal or external finance of R&D is better for company's ongoing. In general, if company finances all R&D expenditures on its own, it will signal its possibility to be innovative for future on its own. But, with external finance the amount of money can be bigger and there can be more output from R&D department. For investors the amount of internal finance in comparison to company's peer is a good indicator to evaluate company's future.[127]

R&D expenditures are linked to company's market position. Companies with high R&D expenditures are the innovative leaders within the business area. So, they are the first mover in connection to new products. First movers are the first companies within a market with a new product. That is directly correlated to the corporate identity and to company's prominence to customers and other stakeholders.[128] Therefore, companies with high R&D expenditures and a lot of innovations are well-known to stakeholders and they have a good position in competition.

Companies have their own business strategy. Typically, the business strategy of firm can focus the quality leadership or the price leadership. According to the quality leadership, the amount of patents can be an indicator for the investor whether the company acts in the sense of its business strategy. Quality leadership without leadership in patenting is difficult. Companies can justify their quality leadership just by presenting new products which other firms do not offer. Quality leadership is impossible when the company just copies the ideas of the competitors. In comparison to company's peer group, a decreasing amount of patent applications can be a signal for a decreasing grade of

---

[125]   Cp. Bound, J. et al. (1984), w/o p.
[126]   Cp. Klette, T., Griliches, Z. (2000), p. 376.
[127]   Cp. Schumpeter, J. (1942), w/o p.
[128]   Cp. Klette, T., Griliches, Z. (2000), p. 377.

innovation within the company. It could be an indicator for the investor to end its investment in the company. Otherwise, an increasing number of patent applications can be a signal for new investment in a company.[129]

Especially young firms can profit from involving patents in financial communication. For questions regarding the securitization of a loan, patents can have an important role. Young companies can demonstrate their ability to pay redemption and interests from further cash flows when they can present an invention which is a unique selling proposition (USP) for the company. The USP gives the company the possibility to beat competitors because they are not able to use the invention because of the protection through the patent. The patent is a written document which enables the chance to evaluate in a better way company's future. Without any document, banks and other lenders are not able to take an invention under consideration which is strategic advantage in competition.[130]

Moreover, patents are not just a protection for the company. It is also a protection for the investor or better for his money. If an invention is sufficient protected through patents, the investor has a better basis to calculate its further returns with the investment. The investor can be sure that his company uses the innovation and other competitors are not able to use the invention.[131]

Tying up, patents are important in connection with a planned initial public offering. A high number of patents increase a company's value in the run-up to the initial public offering. This fact is very important for all investors who want to leave the investment through a public offering. Investors have to force the company to increase the amount of patents when they plan to leave the company. With an increasing number of patents, they can increase their returns from the investment.[132]

For increasing the investor's return, it is important that the patent is used fast in practice. The investor can create the highest return on the patent in short time after publishing the patent. More precisely, the investor has to be engaged in the company when the company has applied the patent, but not until it is published to public. Due to this point, a financial communication which includes intellectual properties can improve company's standing at potential new investors. Unpublished intellectual properties are unknown values which are

---

[129] Cp. Wurzer, A. (2004), p. 37.
[130] Cp. Menninger, J., Kunowski, S. (2003), pp. 1180-1182.
[131] Cp. Bigus, J. (2006), p. 940.
[132] Cp. Hsu, D., Ziedonis, R. (2008), pp. 1-6.

very important for investors. Investors can profit from these properties after they are published because after publication the price for the company increases.[133]

The investor has to master the challenge of evaluating a patent. In contrast to the company the investor is not an expert in the business sector. He has often a lack of knowledge and that is why he is not able to judge real value of a patent. Particularly, the evaluation is very difficult in high innovative business sectors like in the field of biotechnology.[134] Therefore, it is important for the investor that he is supported by the company on the one site and on the other site by neutral experts who work in the special business area. At the end, the quality and the quantity of company's patent portfolio is relevant. Investors have to do also an intensive due diligence of the patent portfolio when they plan to acquire a company[135].

Additionally, existing intellectual properties within a target company minimize costs for R&D. Investors minimize their risks when they invest in a company which is technological and procedural up-to-date. R&D costs often are high in company. Furthermore, they are difficult to calculate because there is often no routine within R&D. Innovative firms are state of the art and R&D is not so important for company's short-term development.[136]

Due to the financial communication, companies have to communicate the value of their intellectual properties. Externals have difficulties to differentiate between groundbreaking innovations and innovations which are important but not groundbreaking. With an open communication companies reach more investors.[137]

Patents description has to be short for the patent office. Investors are interested in detailed information to the patent. They need them to evaluate the patent and its contribution to the whole value of the company.[138] Companies have to consider these points. They have to present the patent in a different way as they applied it at the patent office. So, it is important that companies do not just give investors notifications that they applied a patent. They have to explain their inventions to the investors.

---

[133] Cp. Bigus, J. (2006), pp. 939-960.
[134] Cp. Gogoris, A., Clarke, P. (2001), pp. 279-281.
[135] Cp. Restaino, L. (2006), p. 12.
[136] Cp. Edwards, D. (2001), p. 4.
[137] Cp. Graumann, S., Keil, M. (2004), p. 3.
[138] Cp. Hirschey, M. et al. (2001), pp. 223-235.

In general, patents within company's financial communication have several benefits for the company. Companies which have to be innovative to survive in competition can improve their standing disproportionately. Patents demonstrate innovativeness. That point is the reason for securing financing and enhancing reputation through patents in high innovative business areas.[139]

Handling information asymmetries between management and investor is the main aim of financial communication. The management has to confirm its capabilities for increasing the trust in management's competences and trust in management's intention. Patents are innovative. The innovativeness shows management's aim to develop company's products and processes. Furthermore, it confirms these points because patents are external things from a third party.[140]

Patents are visible signs for the investor wherefore its capital is used. Furthermore, it is an indicator for future returns which can be generated through the new patent. Profiting from these positive aspects requires implementation of patents in the financial communication.[141] The achievement of objectives depends on several parameters, for example, company's industry sector.[142]

[139] Cp. Graham, S. et al. (2009), pp. 255-327.
[140] Cp. Hubig, C., Siemoneit, O. (2009), pp. 68-71.
[141] Cp. Grüning, M. (2011), p. 59.
[142] Cp. Chan, H. et al. (1990), pp. 255-276; Chauvin, K., Hirschey, M. (1993), pp. 128-141.

5 Empirical Analysis of Patent's Relevance for Financial Communication

Chapter five includes the empirical analysis of this work. That is why the chapter starts with an introduction in current research as well as in the methodology of this research. At the end, there are different findings due to the hypotheses set up at the beginning.

5.1 State of Research

Several researchers have investigated the influence of R&D on a company's value in the past. Corporate value is directly linked to share price, which is why this research is also important for the question of this work.

Chan et al.'s data set is based on data from 1979 to 1985. They analyzed the announcement effects of R&D for the United States. The researchers found positive announcement effects for high-tech companies. For low-tech firms, they could not confirm their findings.[143]

Chauvin and Hirschey analyzed data from 1988 to 1990. They set the amount of information as well as the R&D expenses in relation to company size. They found stronger R&D activities at large firms. But in general, they found a positive effect of R&D investments for all company sizes. In addition, the efficiency of R&D depends on company's industry sector.[144]

Pinches et al. confirm the findings of Chan et al. as well as Chauvin and Hirschey.[145] Chung et al. could confirm the findings of Chan et al. as well as Chauvin and Hirschey. But they found other important parameters related to R&D. They linked the analyst coverage as well as the outside directors in the board with the R&D investments. They found that the analyst coverage and the number of outside board directors is positive linked with the performance of stock price. An increasing number of analysts and outside board directors lead to positive stock price development.[146]

---

[143] Cp. Chan, H. et al. (1990), pp. 255-276.
[144] Cp. Chauvin, K., Hirschey, M. (1993), pp. 128-141.
[145] Cp. Pinches, G. et al. (1996), 60-69.
[146] Cp. Chung, K. et al. (2003), pp. 161-172.

Furthermore, Booth et al. confirm the positive relations between R&D and share price performance. Additionally to the other findings, they found that the effects are more significant in mature market financed than in bank financed markets.[147]

Stanzel investigated that the patent application has influences on the accuracy of analyst's forecasts. R&D expenses distort the estimation of earnings. The number of patent applications eliminates the inaccuracy in the estimation of earnings and influences analyst's forecasts in a positive way.[148]

5.2   Hypotheses

This work includes an empirical analysis. The analysis is based on the follow-ing hypotheses which have developed from the state of research mentioned before:

> ➢ H1: Patents have a positive significant impact on the short-term share price performance of an SME

> H1 is the general hypothesis which has to be confirmed. H1 is the base for further hypotheses that seek to find a deeper relation between pa-tents and share price development. Finding this relation occurs under consideration of parameters which are important for differentiating be-tween several characteristics of a company.

>> o H2: The main IPC class of a patent is crucial for a positive signifi-cant patent

>> H2 depends on H1. It focuses on the different IPC classes and tries to find a connection between the IPC class and the impact of a patent on share's performance.

>> o H3: Company's industry sector is crucial for a positive significant patent

>> H3 is also based on H1. Analyzing H3 should answer the question whether there are industry sectors that benefit in an extraordinary

---

[147] Cp. Booth, G. et al. (2006), pp. 197-214.
[148] Cp. Stanzel, M. (2007), w/o p.

way when they implement contents related to patents in their financial communication.

o H4: The year of publication is crucial for a positive significant patent

H4 also concentrates on the general findings of H1. It is used to find a relation between patents and the publication year in the form of indicator for crises or other important developments in business.

o H5: The amount of a company's sales is crucial for a positive significant patent

H5 also orients to the findings of H1. It seeks to find dependencies between company's sales and patents. Sales are often a used parameter for differentiating between several companies. That is the reason for investors and financial multipliers that focus on this parameter when looking for new investment possibilities.

o H6: The number of employees is crucial for a positive significant patent

H6 as well as other hypotheses is based on the general findings of H1. H6 focuses on the number of employees within a company. Similar to sales, the number of employees is an important parameter for differentiating between companies. In specific business areas, employees are the deciding parameter for a successful business.

5.3   Data and Methodology

The investigation period extends over ten years. It starts on January 1, 2003, and ends on December 31, 2012. In total, 265 patents from 18 different companies have been analyzed.[149] All companies are small and medium-sized enterprises and they are all listed on a German stock exchange.

The underlying methodology for further analyses is the event study approach. The event study approach is often used to estimate the average information

---

[149]   Cp. Appendix 1: Empirical Analysis.

context. The average information context is estimated on the basis of market reactions, which happens in a little while around the event. The approach used in this work is modeled on the approach of Fama, Fisher, Jensen, and Roll[150], who first analyzed market reactions on stock prices in 1969.

The approach is clearly structured and comprises the following parts. These parts are considered in the next section of this work:[151]

- Determination of the investigation object

- Definition of dataset and investigation period

- Determination of the observation period and the event period

- Choice of a model for calculating abnormal returns

- Choice of a statistical significance test for evaluating the null hypothesis

- Execution of the statistical test

- Analysis and interpretation of results

5.3.1   Data Collection

The data set is based on data from the Bloomberg Professional Service and from the European Patent Office's website.[152] Bloomberg offers all relevant information regarding a company. It is used by plenty of institutional investors and other businesses for gathering quantitative information regarding companies. The European Patent Office has all relevant data pertaining to patents. It is a source of information for national and international people who are interested in patents.

The dates for new patent publications come from European Patent Office's website. Moreover, all information about patents is from this website. The website offers all information to applicants, inventors, and to the patent classifications.

---

[150] Cp. Fama, E. et al. (1969).
[151] Cp. for example Fama, E. et al. (1969) or for a current study Kolari, J., Pynnönen, S. (2010).
[152] Cp. www.epo.org.

Bloomberg provides end-of-day prices for all related shares. The prices exclude dividends and other equity affecting all that is relevant for the price. In the analysis, dividends and other price-relevant elements receive attention. They are added to the price on the payment day.[153]

On basis of the adjusted stock prices, the daily return is calculated by following formula[154]:

$$r_{jt} = \log \frac{p_{jt}}{p_{jt-1}}$$

r_{jt}: One period logarithmic return

p_{jt}: Price of stock j on day t

p_{jt-1}: Price of stock j on day t-1

Formula 1: r_{jt}: One period logarithmic return

Log returns are mutually independent. The continuous return offers in contrast to the historical return the opportunity to calculate average returns on the basis of the arithmetic mean. Besides, it is possible to add up each return. Log returns are associated with the geometric random walk model which supports the idea of efficient market hypothesis. Efficient market hypotheses are related to all valuable information. All information has impacts on the development of a stock.[155]

The later used market return r_{mt} is calculated in same way as r_{jt}.

In the following section, the criteria for choosing an event are explained.

5.3.2   Event Study Approach

The event study approach has a long history. During this time, the approach has not changed much. The approach was first used in the 1930s. The cur-

---

[153]   Cp. Appendix 1: Dividends_var and Final Dataset.
[154]   Cp. Ruppert, D. (2004), pp. 96-97.
[155]   Cp. Ruppert, D. (2004), pp. 96-97.

rently used approach[156] is based on Ball and Brown (1968) and Fama et al. (1969).

Event studies are often considered in financial and capital market research. The reason for this is the adjustment of this approach to several investigation objects. Economic as well as legal objectives can be investigated with the event study approach. More than 500 articles related to the event study approach were published during the period 1974–2000.[157]

On the basis of an empirical statistical tool in the research field of accounting and finance, the event study approach has been assimilated into other disciplines. The approach is used in economics, history, law, and political sciences. It is easy to adapt to other questions in research. The core idea of the approach is easy to fit to new variations and methodologies.[158] In general, it supports valuing new information with regard to the stock of a company.

The event study approach is adjustable. New findings and new technological developments are suitable for the approach. Furthermore, the approach includes relevant findings of the financial research. Examples are the included capital asset pricing model (CAPM) of Sharpe (1964) and the general possibility to verified elements with an empirical tool. Initially, capital structure issues could be investigated.[159] [160]

Early event studies use monthly data for analyses.[161] Monthly data is inaccurate. Inaccuracies falsify the final result. The final result becomes influenced by events other than the investigated one. During a monthly period, there are different happenings which could impact stock price development.

The event is described as the day when the investigated issue is announced. Usually, the event is investigated with the addition of some trading days before the event and some days after the announced event date.[162] The event happens in a short period of time. If the period is too long, there will be other events which have an impact on the performance. Analyzing one event im-

---

[156]   For current research, check, for example, Cong, Hoitash, and Krishnam (2010), as well as Kolari and Pynnönen (2010).
[157]   Cp. Kothari, S., Warner, J. (2007), pp. 3-36.
[158]   Cp. Corrado, C. (2011), p. 207.
[159]   Relevant papers in this context are Modigliani and Miller (1958) as well as Miller and Modigliani (1961 and 1963).
[160]   Cp. Corrado, C. (2011), pp. 207-208.
[161]   Cp. e.g. Brown, S., Warner, J. (1980); Shevlin, T. (1981).
[162]   Cp. Corrado, C. (2011), p. 209.

pacts the development of a stock. As a result, there must be a clearly differentiation between other events.

There was a quick change to daily data, when they were available.[163] Daily data has a closer relationship to one special event. The event can be investigated separately from other events by using daily data. Daily data helps to clearly differentiate one event from another.

The approach also contains some critical elements which have to be mentioned using the approach. To begin with, the information, on which the event is based, has to be absolutely new. Regarding the efficiency of the capital markets, all information is implemented in the stock price. If the information is old, it is already in the price and the analysis does not work. In addition, the approach only provides a general answer to the question whether or not there is a change in the stock price. The approach does not answer the question whether the price increases or decreases. For answering this question, further analyses are necessary.[164]

5.3.3 *Criteria for Choosing an Event*

The basic population for the succeeding empirical analysis is based on patents that fulfill all of the following criteria:

- Patent publication happened in the period between January 1, 2003, and December 31, 2012,

- Companies have to be a current member of the Prime[165] Standard at the Frankfurt Stock Exchange,[166]

- Companies have to be a small or medium-sized enterprise in line with the EU definition for SMEs,

- Daily end-of-day prices have to be provided by Bloomberg for the observation period and the event period,

---

[163] Cp. Brown, S., Warner, J. (1985).
[164] Cp. Agrawal, J., Kamakura, W. (1995), pp. 56-62; Clement, M., Fischer, M., Goerke, B. (2007), pp. 418-444; McWilliams, A., Siegel, D. (1997), pp. 626-657.
[165] For further details compare Deutsche Börse (2012).
[166] Classically, SMEs are not listed at a stock exchange. For using the event approach as research method, it is necessary that the companies are listed and that there are historical prices for companies' shares.

- The time lag between each patent publication of a company has to be at least 30 days.

After preparing the final data set, there are 265 patents from 18 different companies. All 265 patents satisfy the criteria defined earlier.

A plenty of companies apply for a lot of patents. Quite often, the application rate is higher than two patents per publication date. These companies are considered once.

The prime standard at the Frankfurt Stock Exchange is regulated on EU criteria. This high trustworthiness increases popularity for the stock members for national and international investors. That is why the trade volume of shares is higher than in other standards. Stock members have to satisfy the highest level of transparency requirements. The transparency requirements in the prime standard are higher than the legal statutory provisions. More than the legal statutory provisions, stock members have to report quarterly in German and English. They are required to maintain international reporting standards (IFRS or US-GAAP). Events have to be communicated in a company's calendar. They have to manage a yearly conference for analysts. Additionally, they have to publish ad hoc announcements in the English language. At the end, the prime standard is a condition for getting membership of the DAX, MDAX, TecDAX, or SDAX[167].[168]

The investigation period includes maximum five days before the patent publication takes place and five trading days after the event [-5; 5]. For detailed information, following periods are investigated: [-3; 0], [-1; 0], [0; 1], and [0; 3]. The comparable values from the observation period build up for the 20 trading days before patent publication: [-20; -6]. In other studies, the observation period is often longer. In this special case, a longer observation period is not possible to define. A lot of companies often apply for patents. A longer observation period poses the risk that the events are not clearly differentiated, which is also the reason for choosing the time lag of 30 days between all patents. Without choosing a time lag, it is impossible to generate a logical connection between applying for patent A and patent B. To reach good findings, it is important to strictly separate all events.

---

[167]   The DAX, MDAX, TecDAX, and SDAX are the most common German indices for stocks.
[168]   Cp. Deutsche Börse (2012), pp. 1-4.

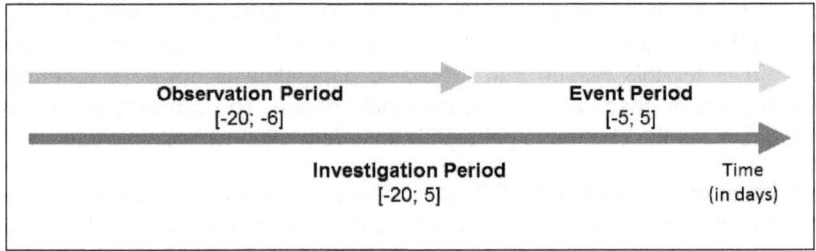

Figure 6: Investigation Period
Own graphic.

5.3.4 Descriptive Statistics

The basic population of this work comprises 265 patents from 18 different companies. The underlying patents are applied between January 1, 2003, and December 31, 2012. The following graphic gives a chronological overview of the number of published patents per year:

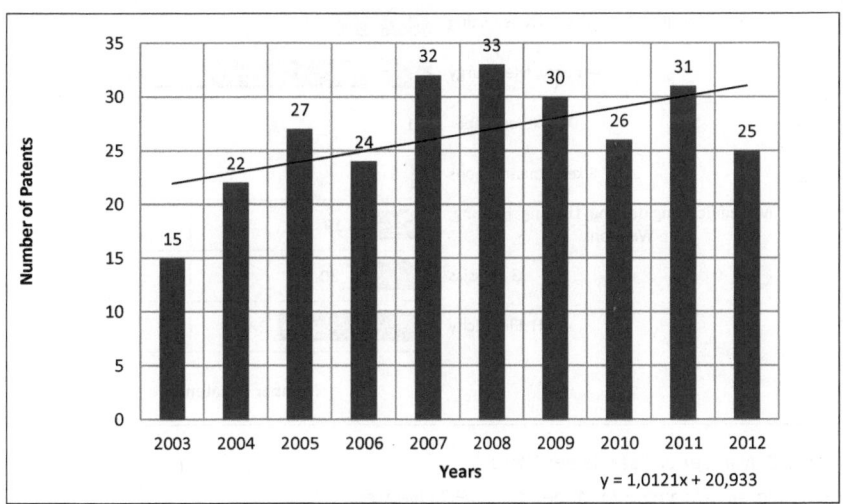

Figure 7: Number of Patent Publications per Year
Own diagram according to: Appendix 1, Number of Publications per Year.

The number of patent publications during the investigation period rises slightly. The graph indicates a slope of 1,0121 additional patent publications per year.[169] It is for this reason that the basic population is not so staggering. There are similar observations over the whole period. The similarity in the observations gives the chance to get significant information without any outliers.

The number of patents in one IPC class differs from class to class. The class with the highest number of patent publications is Class A. Class A includes 83 patents that deal with human necessities. It is followed by Class C. 69 Class C patents deal with chemistry and metallurgy. There are 38 patents in Class H. All these patents are related to electricity. Class F (mechanical engineering, lighting, heating, and weaponry) and Class G (physics) are on the same level with 19 and 20 patents respectively. Less important for further analyses are Class D for textiles and paper, as well as Class E for fixed constructions, because the number of patents remains far too low to deal with it in an adequate way.[170]

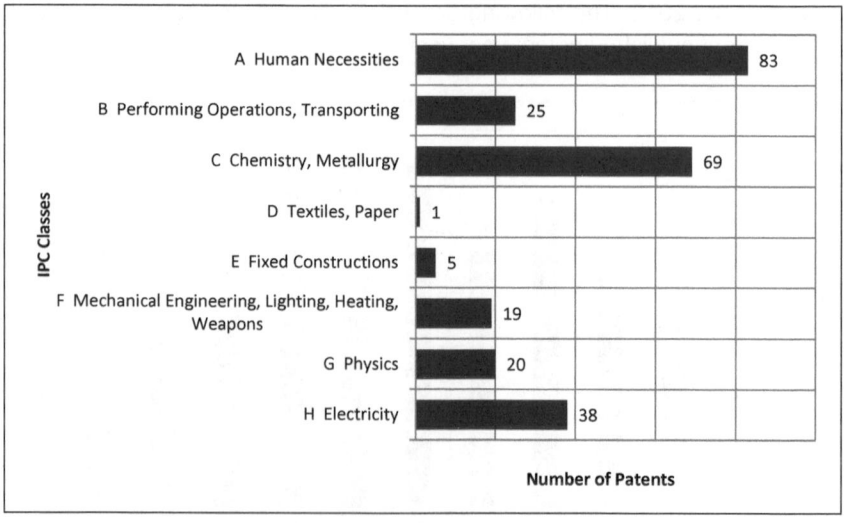

Figure 8: Number of Patents per Class
Own diagram according to: Appendix 1, IPC Overview.

---

[169] Cp. Figure 7: Number of Patent Publications per Year.
[170] Cp. Figure 8: Number of Patents per Class.

The basic population consists of 18 companies. The companies are from different industry sectors. The most widely represented sector is consumer goods.[171] This sector also boasts the highest rate of patent publications with 160 patents in the investigation period. The sector for consumer goods is followed by the technology sector with seven companies and 10 patent publications. The number of companies in the energy and in the industrial sector is on the same level. There are three companies in both sectors. The number of patents differs in both sectors. In the energy sector there are 46 patent publications and in the industrial sector there are 18. Only one company belongs to the communication sector. This company has applied for 31 patents during the investigation period.[172]

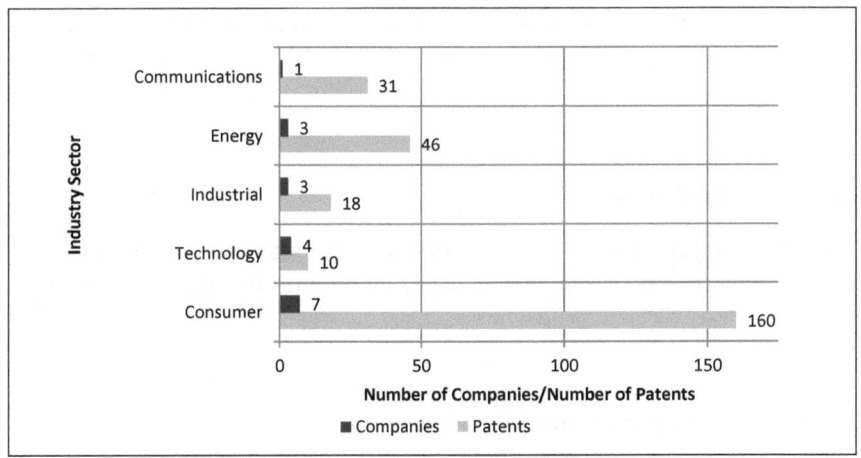

Figure 9: Number of Companies and Patents per Industry Sector
Own diagram according to: Appendix 1, Industry Sector.

In general, the data set is comparable. There are some observations which have to be looked at closely. To begin with, it is the patent Classes D and E. Both classes have fewer values.[173] Hence, the results can affect further investigations. Owing to the same reason, the industry sector communications is not representative.

---

[171]   Data is based on Bloomberg. Bloomberg differs in terms of non-cyclical and cyclical consumer goods. This investigation does not differentiate between these sectors because of simplification.
[172]   Cp. Figure 9: Number of Companies and Patents per Industry Sector.
[173]   Cp. Figure 8: Number of Patents per Class.

5.3.5 Calculation of Abnormal Returns

The abnormal return is the difference between real and expected returns. If it is positive, it is an outperformance; otherwise, it is an underperformance. There are different ways of calculating the abnormal return. The abnormal return has to consider market conditions as well as firm-specific conditions. In the following, several methods for calculating the abnormal return are mentioned. The one used in this paper is explained in detail:[174]

1) Mean-adjusted returns

The model using mean-adjusted returns rests ex ante on constant expected returns on security j. Whether the expected return is constant varies from company to company.

2) Market-adjusted returns

The market-adjusted returns need equal expected returns for all securities ex ante. In practice, it is difficult that all expected returns have the same expected return.

3) Capital asset pricing model benchmark

In practice, the CAPM is often used to calculate the abnormal return on a security. It is for this reason that the CAPM considers the security risk as well as the market risk. The core idea of the CAPM is not adjustable to all investigations.

4) Market model benchmark

The general approach is the market model from Sharpe (1963). It is the model used in this paper. The market model does not make explicit assumptions about equilibrium stock prices. This model calculates expected returns in the following way:

---

[174] Cp. Strong, N. (1992), pp. 536-538.

$$R_{jt} = \alpha_j + \beta_j R_{mt} + u_{jt}$$

R_{jt}: Expected Return of Security j on Day t

α_j; β_j: Estimation parameters of Linear Regression of Stock j on Day t

R_{mt}: Return of Stock Market on Day t

u_{jt}: Residuum of Linear Regression

Formula 2: R_{jt}: Expected Return of Security j on Day t

In the model u_{jt} is uncorrelated with R_{mt}. R_{mt} is a component which reflects the systematic risk. u_{jt} is the unsystematic component. The unsystematic component captures all firm-specific events. Therefore, the information signal and R_{mt} have to be independent. Considering these aspects u_{jt} is zero without any expectations and with an independent variance of t. α_j and β_j have to be estimated for each stock.[175]

α_j and β_j are firm-specific constants. α_j is the intercept. The intercept measures the outperformance of stock j. The outperformance is adjusted by market risks. β_j is the slope. The slope shows the correlation between security j and benchmark index R_{mt}.

After calculating the expected return on security j on day t, the abnormal return on security j on day t is calculated. The abnormal return describes the unexpected return on security j on day t. The unexpected return results from the investigated event, in this case the patent publication. The abnormal return is calculated in the following way:[176]

---

[175] Cp. Sharpe, W. (1963), p. 281.
[176] Cp. Cheng, F., Ariff, M. (2007), p. 5.

$$AR_{jt} = r_{jt} - [\alpha_j + \beta_j R_{mt}] = r_{jt} - R_{jt}$$

AR_{jt}: Abnormal return of security j on day t

r_{jt}: Logarithmized return of security j on day t

α_j; β_j: Estimation parameters of linear regression of stock j on day t

R_{mt}: Return of stock market on day t

R_{jt}: Expected return of security j on day t

Formula 3: AR_{jt}: Abnormal return of security j on day t

The market model leads to more powerful statistical tests. It results in smaller variances of abnormal returns in connection with raw returns. The level of correlations across secure abnormal returns is smaller, which gives greater conformity to standard statistical tests.[177]

Basis for the abnormal return is the logarithmized raw return. It allows summing up daily returns to get an overview of the abnormal returns of one period.

5.3.6 Significance Tests

For verifying the assumptions that patent publications impact share price development, it has to be tested whether the abnormal return differs significantly from null on the event day respectively in the event period around the event day. Therefore, the null hypothesis and the alternative hypothesis are as follows:

H_0: Abnormal returns do not differ significantly from null on the event day,

H_A: Abnormal returns differ significantly from null on the event day.

---

[177]   Cp. Beaver, W. (1968), pp. 67-92.

$$t_0 = t_{emp} = \frac{AR_{jt}}{\sigma_{AR_{jt}}}$$

$t_0 = t_{emp}$: Empirical t-value

AR_{jt}: Abnormal return of security j on day t

$\sigma_{AR_{jt}}$: Standard deviation of AR_{jt}

Formula 4: $t_0 = t_{emp}$: Empirical t-value (Abnormal return)

The event period is investigated using accumulated abnormal returns. The hypotheses for the event period are related to the mentioned ones:

H_0: Accumulated abnormal returns do not differ significantly from null during the event period,

H_A: Accumulated abnormal returns differ significantly from null during the event period.

$$t_0 = t_{emp} = \frac{1}{n} \sum_{j=1}^{n} \frac{AAR_{jt}}{\sigma_{AAR_{jt}}}$$

$t_0 = t_{emp}$: Empirical t-value

AAR_{jt}: Accumulated abnormal return of security j on day t

$\sigma_{AAR_{jt}}$: Standard deviation of AAR_{jt}

n: Number of events in one period

Formula 5: $t_0 = t_{emp}$: Empirical t-value (Accumulated abnormal return)

Due to the distribution of t, the significance can be taken from the relative tables.[178] The average abnormal return of a security is required for executing the significance tests. It is calculated in the following manner:

---

[178]   Cp. Brown, S., Warner, J. (1980), pp. 205-258.

$$\overline{R_{jt\,n,[t_0,t_1]}} = \frac{1}{N}\sum_{n=1}^{N}\overline{R_{jt}}$$

n,t

$\overline{R_{jt\,n,[t_0,t_1]}}$: Average abnormal return of a security in time t

N: Number of abnormal returns of security j in one period with n=1, ..., N

$\overline{R_{jt}}_{n,t}$: Abnormal return of security j in time t

Formula 6: $\overline{R_{jt\,n,[t_0,t_1]}}$: Average abnormal return of a security in time t

5.4   Results of Empirical Analysis

The empirical analysis confirms the link between patents and stock price performance. Intellectual property rights have significant impacts on SME's stock price performance.

The empirical analysis was done with a short-term focus. Regarding the described data and methodology, the empirical evidence was only investigated for small and medium-sized enterprises.

In all, the data set includes 265 patents from 18 small and medium-sized companies. The SMEs are listed on a stock exchange and they fulfill all characteristics of the EU definition of SMEs. The analyzed companies differentiate in terms of sales, number of employees, and industry sector. That is the reason why there is firstly a general analysis of the whole data set. The general analysis is followed by specific analysis of different characteristics.

The findings confirm Hypothesis 1: Patents have a positively significant impact on the short-term share price performance of an SME. Table 3 shows that 173 out of 265 analyzed patents have significant data in one of the respective investigation periods. Sometimes, there are results for particular days and sometimes there are results for particular investigation periods. In general, there are 468 significant days or investigation periods in the whole data set.

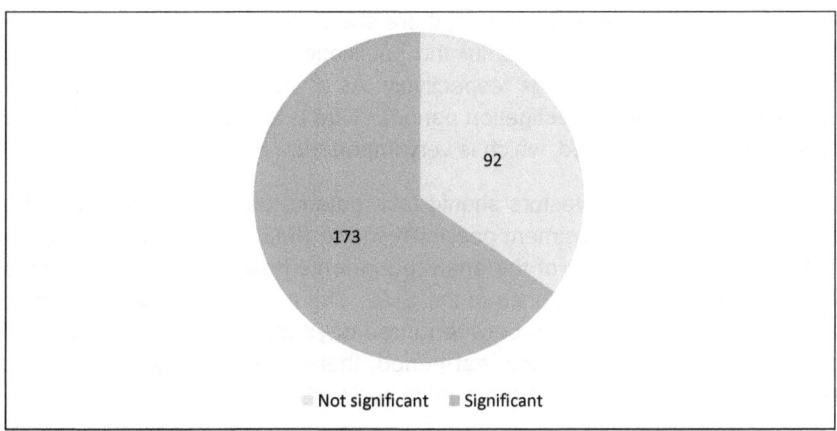

Figure 10: Significance of Investigated Patents
Own diagram according to: Appendix 1

The significance is tested on a level of significance of 99%, 95%, and 90%. Under consideration of the whole data set, most findings are evident on a level of 90%. 177 out of 468 findings are evident on a 90% level of significance. It is followed by 161 out of 468 on 95% and 130 out of 468 on a 99% level of significance.

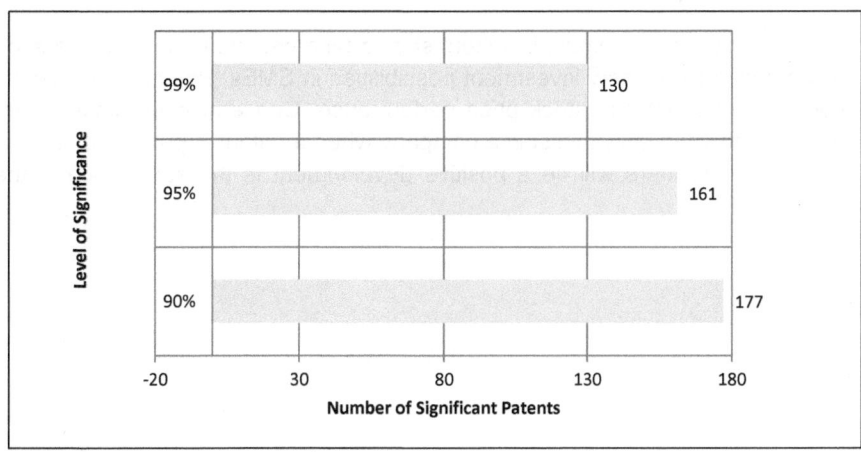

Figure 11: Overview of Level of Significance per Days
Own diagram according to: Appendix 1.

Prior to the publication of the patent, there are also findings as well as after the publication. Five and four days for the publication date, there are 39 and 38 out of 468 significant findings respectively. As a whole, there is less deviation between all days and investigation periods. That is a reason why there can be no specific day mentioned, which is very important.

Concerning Table 3, investors should take patents under consideration when they look for a new investment possibility in the SME sector from investigated companies. 65% percent of the analyzed patents have a significant influence on the share price performance of the SME. The most important period for investors with most findings is between three days for the publication date until the date of publication. During that period, there is the strongest relation between patent publication and share price performance.

This observation suggests that some people check patent publications when they invest in an SME. It seems that some people get information before the publication date because strongest changes in the stock price are from three days before the date of publication until the date of publication.

These findings support those of Chauvin and Hirschey. Chauvin and Hirschey as well as these findings confirm that there is a significant relation between R&D and company's general performance. In this work, patents are the related factor instead of R&D in general.[179]

Considering these findings, investors should take patents under consideration when they look for new investment possibilities in SMEs. Patents have significant relations with the stock price performance. Concerning the findings, investors should directly invest in a company when a patent is published. In 65% of all patents, there will be a positive development in the stock price in the short-run.

---

[179]   Cp. Chauvin, K., Hirschey, M. (1993), pp. 128-141.

	Total	t-5	t-4	t-3	t-2	t-1	t	t+1	t+2	t+3	t+4	t+5	[-5; 5]	[-3; 0]	[-1; 0]	[0; 1]	[0; 3]	Total Significance
Significant	173	39	38	32	38	31	28	38	28	31	28	37	15	22	22	21	20	468
Total	265																	
Percentage	65%																	
99%	173	11	12	11	16	8	6	11	6	9	8	10	3	3	6	5	5	130
95%	173	17	15	13	10	8	13	12	9	8	10	13	3	10	4	10	6	161
90%	173	11	11	8	12	15	9	15	13	14	10	14	9	9	12	6	9	177
		39	38	32	38	31	28	38	28	31	28	37	15	22	22	21	20	468
		8,33%	8,12%	6,84%	8,12%	6,62%	5,98%	8,12%	5,98%	6,62%	5,98%	7,91%	3,21%	4,70%	4,70%	4,49%	4,27%	

Table 3: Overview of Total Empirical Analysis

In the following sections, the general findings of the whole data set will be analyzed by dividing the whole data set into several categories.

5.4.1 Results of Empirical Analysis per International Patent Classification

In the previous section, it was confirmed that there is a general dependency between the patent application and the stock price performance of a small and medium-sized listed company. Patents have a positive impact on this performance.

The data delivers further findings when they are divided by the IPC of each patent. All succeeding data is shown in Table 4. Concerning the IPC, there are 83 patents in Class A, 25 patents in Class B, 69 patens in Class C, one patent in Class D, five patents in Class E, 19 patents in Class F, 20 patents in Class G, and 38 in Class H.

55 out of the 83 patents in Class A, 15 of the 25 patents in Class B, 48 out of 69 patents in Class C, one out of one patent in Class E, five out of five patents in Class E, 12 out of 19 patents in Class F, seven out of 20 patents in Class G, and 28 out of 38 patents in Class H have significant data in the investigated period of time. According to the limited amount of patents in Classes E and F, the following analyses concentrate on the other IPC classes.

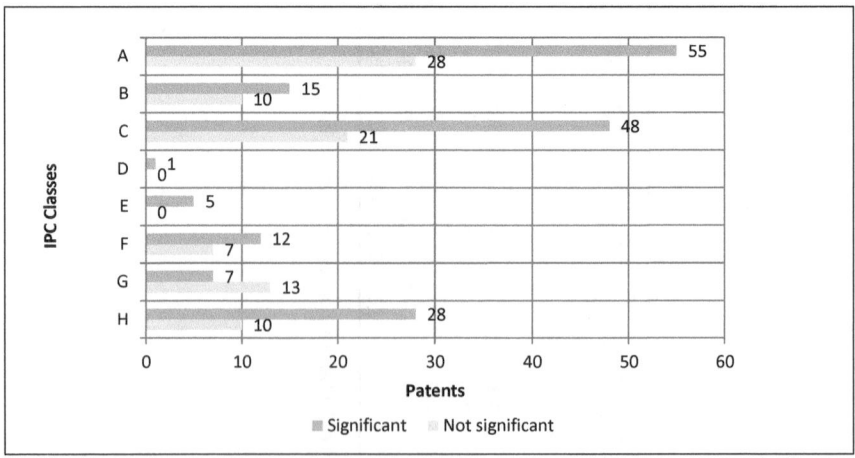

Figure 12: Significance per IPC Class
Own diagram according to: Appendix 1

The IPC class is based on the official patent publication. Defining an IPC class happens in this case according to the most mentioned IPC class in the official patent publication from the European Patent Office. A great number of patents belong to different IPC classes.

Most significant patents are in Class H. In Class H, 28 out of 38 patents have a significant relation to shares performance. Class H includes patents that deal with electricity. There are 74% of all patents within Class H are significant.

After Class H, Class C includes the highest amount of significant patents. 48 of 69 patents or 70% of all patents in this class are significant. Class C covers patents dealing with chemistry and metallurgy.

In Class A, 55 out of 83 patents are significant. This is 66% of all patents in this class. Class A covers patents which focus on human necessities. Patents from Class F with 12 out of 19 patents or 63% of all patents in this class are significant. In Class F are all patents that focus on mechanical engineering, lighting, heating, and weapons.

According to the publication date, the general trend that the most interesting period of time is some days before the publication cannot be confirmed for all IPC classes. Classes B, C, and H include the most patents which are significant in the run-up to the publication date. All other classes have significant patents before and after the publication date. In particular, there is no large deviation between the classes. Concerning these findings, a general statement about which class is particularly interesting for an investor in the run-up or in the follow-up is not possible.

To summarize the findings after analyzing the different IPC classes, investors should focus on patents from all IPC classes. Apart from Class G, in all classes are greater than or equal 60% of patents are significant. The most interesting IPC class in this connection is Class B. In this class, the number of total significant days or period of times is 50. Relating to the number of patents, each patent has 3,33 significant days or period of times. Class B is followed by Class H with 2,79 and Class C with 2,73 days or period of times. If investors take the IPC under consideration, they should focus on Classes B and H. In these classes, the number of significant patents and the biggest amount of significant days as well as the time periods can be found.

These findings suggest that especially inventions that improve performing operations and transportation and those inventions being technological inventions in the field of electricity are very important for investors and finally for a

Summary

	A	B	C	D	E	F	G	H
Significant	55	15	48	1	5	12	7	28
Total	83	25	69	1	5	19	20	38
Percentage	66%	60%	70%	100%	100%	63%	35%	74%

Overview of Empirical Analysis per IPC (Total Significance = 465)

IPC		t-6	t-5	t-4	t-3	t-2	t-1	t	t+1	t+2	t+3	t+4	t+6	[-6; 5]	[-3; 0]	[-1; 0]	[0; 1]	[0; 3]	Total Significance
A	99%	2	4	4	3	8	3	3	4	2	3	2	2	1	1	2	3	2	45
A	95%	8	2	2	2	3	2	4	3	4	2	4	4	0	4	1	3	1	47
A	90%	3	5	5	3	4	6	1	5	2	6	4	3	2	2	5	1	4	57
A	**Total**	**13**	**11**	**11**	**8**	**15**	**11**	**8**	**12**	**8**	**11**	**10**	**9**	**3**	**7**	**8**	**7**	**7**	**149**
A	%	8,72%	7,38%	7,38%	5,37%	10,07%	7,38%	5,37%	8,05%	6,04%	7,38%	6,71%	6,04%	2,01%	4,70%	5,37%	4,70%	4,70%	
B	99%																		10
B	95%																		22
B	90%																		18
B	**Total**	**4**	**5**	**5**	**5**	**2**	**2**	**4**	**3**	**4**	**3**	**0**	**4**	**3**	**3**	**3**	**2**	**3**	**50**
B	%	8,00%	10,00%	10,00%	10,00%	4,00%	4,00%	8,00%	6,00%	8,00%	6,00%	0,00%	8,00%	6,00%	6,00%	6,00%	4,00%	6,00%	
C	99%																		31
C	95%																		50
C	90%																		50
C	**Total**	**9**	**11**	**7**	**7**	**10**	**10**	**8**	**9**	**4**	**8**	**12**	**14**	**4**	**7**	**7**	**7**	**5**	**131**
C	%	6,87%	8,40%	5,34%	5,34%	7,63%	7,63%	6,11%	6,87%	3,05%	6,11%	9,16%	10,69%	3,05%	5,34%	5,34%	5,34%	3,82%	
D	99%																		1
D	95%																		1
D	90%																		0
D	**Total**			1									1						**2**
D	%	0,00%	0,00%	50,00%	0,00%	0,00%	0,00%	0,00%	0,00%	0,00%	0,00%	0,00%	50,00%	0,00%	0,00%	0,00%	0,00%	0,00%	
E	99%																		5
E	95%																		5
E	90%																		8
E	**Total**	**0**	**1**	**1**	**0**	**0**	**1**	**1**	**1**	**0**	**1**	**0**	**1**	**0**	**0**	**1**	**1**	**0**	**18**
E	%	0,00%	5,56%	5,56%	0,00%	0,00%	0,00%	0,00%	0,00%	0,00%	5,56%	0,00%	5,56%	50,00%	0,00%	5,56%	0,00%	0,00%	
F	99%																		4
F	95%																		12
F	90%																		7
F	**Total**	**3**	**1**	**1**	**2**	**1**	**2**	**2**	**2**	**1**	**1**	**0**	**2**	**0**	**0**	**1**	**0**	**1**	**23**
F	%	13,04%	5,56%	5,56%	11,11%	4,35%	8,70%	11,11%	11,11%	13,04%	5,56%	0,00%	8,70%	0,00%	0,00%	4,35%	0,00%	5,56%	
G	99%																		4
G	95%																		6
G	90%																		4
G	**Total**	**1**	**1**	**0**	**1**	**1**	**0**	**3**	**1**	**1**	**0**	**0**	**1**	**0**	**1**	**0**	**0**	**0**	**14**
G	%	7,14%	7,14%	0,00%	14,29%	21,43%	0,00%	21,43%	7,14%	7,14%	0,00%	0,00%	7,14%	0,00%	7,14%	0,00%	0,00%	0,00%	
H	99%																		30
H	95%																		18
H	90%																		30
H	**Total**	**8**	**5**	**5**	**5**	**5**	**5**	**2**	**7**	**6**	**8**	**4**	**6**	**4**	**5**	**2**	**2**	**4**	**78**
H	%	10,26%	6,41%	6,41%	6,41%	6,41%	6,41%	2,56%	8,97%	7,69%	10,26%	5,13%	7,69%	5,13%	6,41%	2,56%	2,56%	5,13%	

Table 4: Overview of Empirical Analysis per IPC

company's value. That is a logical thing because all inventions due to performing operations and transportation decrease costs for the company. Decreasing costs support company's role in the competition. Inventions of highly technical nature can be very valuable in the future for a company's role in competition. According to Hypothesis 2, the main IPC class of a patent is crucial for a positive significant patent. The hypothesis can be confirmed in some parts but the findings are not sufficient to give a general suggestion for dealing with the IPC class.

5.4.2 Results of Empirical Analysis per Industry Sector

The empirical analysis includes 18 companies from different industry sectors. These sectors are communications, consumer (cyclical), consumer (non-cyclical), energy, industrial, and technology.

Hypothesis 3 posits: Company's industry sector is crucial for a positive significant patent can be confirmed with the findings. Owing to the industry sectors, the industry sector energy has the largest number of significant patents with eight out of 10 or 80% of all patents. It is followed by the industry sector communications with 24 out of 31 patents or 77% of all patents. It is followed by consumer (cyclical) with six out of eight patents, or 75%. The industry sector consumer (non-cyclical) has a total of 152 patents. 103 patents are significant and this is equal to 68% of all patents within the sector. Industrial and technology are less important to analyze the significance in each industry sector. In sector technology, only 25 out of 46 patents or 54% are significant. Seven out of 18 patents are significant in the industrial sector. This is equal to 39% of all patents.

Because of the industry sectors, the analysis shows that there are larger deviations between the industry sectors than in the previous findings. These findings support the findings from Chauvin and Hirschey.[180] Owing to the influence of R&D on the whole value of a company, there are differences between the industry sectors.

---

[180] Cp. Chauvin, K., Hirschey, M. (1993), pp. 128-141.

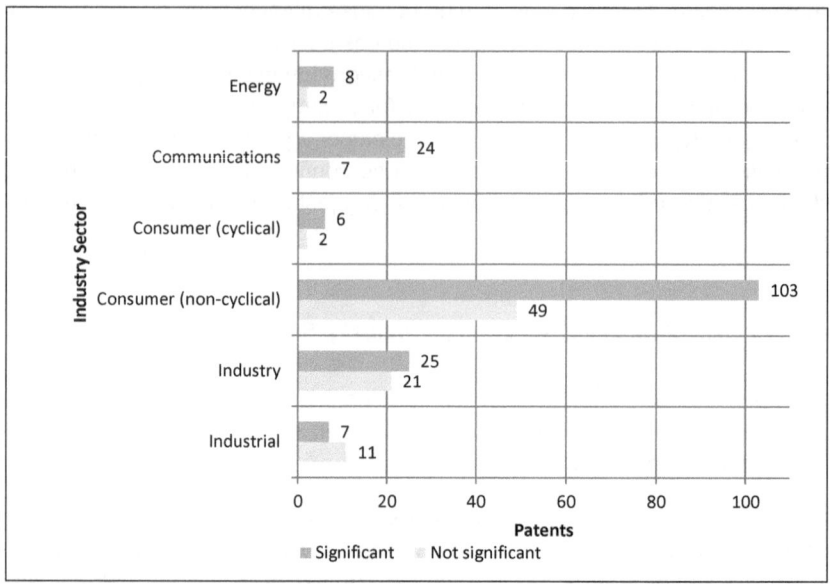

Figure 13: Significance per Industry Sector
Own diagram according to: Appendix 1

The findings agree with the point that some industry sectors need more inno-vativeness than others. The energy sector is very cost intensive.[181] In the en-ergy sector, shareholders benefit directly from technological progress because the inventions decrease costs. This industry sector has a long-term focus, which is the reason for making extraordinary profit in comparison with other sectors. There is plenty of capital invested in production in the energy sector. All inventions improving the production decrease the costs for production sig-nificantly. The communication sector needs innovation in another way which is different from the energy sector to survive in competition. The communication sector needs inventions to be up to date.[182] The product lifecycle is very short in the communication sector and companies need invention to launch new products with a lot of new technological features in the market. All the men-tioned features lead to stronger dependency of these industry sectors on new inventions. That is why investors concentrate on patents when looking for new investment possibilities in the fields of energy and communication.

---

[181]   Cp. Bashmakov, I. (2009), pp. 54-57.
[182]   Cp. Goto, A. (2009), pp. 58-59.

The findings from the energy sector and the communication sector can be transferred to the consumer sector, both cyclical and non-cyclical. Here, customers constantly demand new products, leading to a short product life cycle in this sector. Consumers often change their needs and companies have to respond to consumers' needs.

Surprisingly, the technology sector does not have such a large number of significant patents. This finding supports the findings of Chan, H. et al.[183] There cannot be seen a general impact of innovativeness on companies from the technology sector.

Besides, there is a low impact of innovativeness on the performance of company's share price in the industrial sector. In this sector, other factors are more important than innovativeness. Companies belonging to this sector often use their engines for the long term and do not need a lot of technological innovativeness. They have to improve their productivity based on other factors like better buying conditions or rationalization.

Based on the findings in this part, there is no suggestion for investors to invest on a special day in a specific share. There are also findings before and after the publication date.

Most significant findings are in the industry sector consumer (cyclical). There are 28 findings with only six patents. This leads to the assumption that companies from the consumer sector have to communicate in the open when the have a new patent. This can lead to a better share price performance. This point is supported by the consumer (non-cyclical) industry sector. There is also with a factor of 2,57 a lot of significant patents within the whole number of analyzed patents.

Considering the whole gamut of significant findings, the previously mentioned point that the communication sector needs a lot of inventions to be competitive can be confirmed. There are significant findings for each patent. This cannot be confirmed for the energy sector.

In general, patents are very important for several industry sectors. Obviously, the communication sector is much dependent on innovativeness. Therefore, it is important for companies from that sector to embed patents and other facts of innovativeness in the financial communication. Investors take these things under consideration when they look for new investment possibilities in this

---

[183] Cp. Chan, H. et al. (1990), pp. 255-276.

	Communications	Consumer, Cyclical	Consumer, Non-cyclical	Energy	Industrial	Technology
Significant	24	6	103	8	7	25
Total	31	8	152	10	18	46
Percentage	77%	75%	68%	80%	39%	54%

	t-5	t-4	t-3	t-2	t-1	t	t+1	t+2	t+3	t+4	t+5	[-5;5]	[-3;0]	[-1;0]	[0;1]	[0;3]	Total Significant Periods	Significant Periods
period count	39	38	32	38	31	28	38	28	31	28	37	15	22	22	21	20	468	
Communications																		
99%	3	2	3	2	1	0	2	3	3	2	1	2	1	1	1	3	30	1,25
95%	1	1	2	1	1	0	2	1	2	0	0	1	3	1	1	0	17	0,71
90%	2	1	0	2	2	2	2	3	2	2	4	1	1	0	0	1	25	1,04
Total	6	4	5	5	4	2	6	7	7	4	5	4	5	2	2	4	72	3,00
%	8,33%	5,56%	6,94%	6,94%	5,56%	2,78%	8,33%	9,72%	5,56%	5,56%	6,94%	5,56%	6,94%	2,78%	2,78%	5,56%		
Consumer, Cyclical																		
99%	0	0	1	0	0	0	0	1	0	0	1	0	0	0	0	0	6	1,00
95%	3	1	1	0	0	3	1	1	0	0	0	2	0	0	2	2	15	2,50
90%	0	0	1	0	0	0	0	0	0	1	2	0	2	1	2	0	7	1,17
Total	3	2	1	2	0	3	2	2	2	1	2	2	2	1	2	2	28	4,67
%	10,71%	7,14%	3,57%	7,14%	0,00%	10,71%	7,14%	3,57%	7,14%	3,57%	7,14%	7,14%	7,14%	3,57%	7,14%	7,14%		
Consumer, Non-cyclical																		
99%	7	6	4	12	3	9	7	2	4	6	4	1	2	2	4	2	69	0,67
95%	9	8	7	6	5	9	5	5	4	8	10	2	4	2	6	3	95	0,92
90%	6	9	4	7	10	4	9	5	9	6	8	3	4	9	2	6	101	0,98
Total	22	23	15	25	18	16	23	12	17	20	22	6	10	13	12	11	265	2,57
%	8,30%	8,68%	5,66%	9,43%	6,79%	6,04%	8,68%	4,53%	6,42%	7,55%	8,30%	2,26%	3,77%	4,91%	4,53%	4,15%		
Energy																		
99%	0	0	0	0	0	1	0	0	0	0	1	0	0	1	0	0	4	0,50
95%	0	0	0	0	0	1	0	1	0	1	0	0	0	0	0	0	3	0,38
90%	1	0	1	0	0	0	0	0	1	0	1	0	1	0	1	0	7	0,88
Total	1	0	1	0	2	1	0	1	1	1	2	0	1	1	1	1	14	1,75
%	7,14%	0,00%	7,14%	0,00%	14,29%	7,14%	7,14%	7,14%	7,14%	7,14%	14,29%	0,00%	7,14%	7,14%	7,14%	0,00%		
Industrial																		
99%	1	2	0	0	1	0	0	0	0	0	0	0	0	0	0	0	3	0,43
95%	0	2	0	0	0	0	0	0	0	0	2	0	0	0	0	0	4	0,57
90%	0	0	1	0	2	0	1	1	0	1	0	0	3	2	0	1	5	0,71
Total	1	2	1	0	2	0	1	1	1	0	2	0	0	2	0	1	12	1,71
%	8,33%	16,67%	8,33%	0,00%	16,67%	0,00%	8,33%	8,33%	8,33%	0,00%	16,67%	0,00%	0,00%	0,00%	0,00%	8,33%		
Technology	25																	
99%	1	3	4	1	3	2	1	1	0	1	0	0	2	0	1	0	18	0,72
95%	3	3	3	3	1	1	1	1	2	2	2	0	3	1	1	1	27	1,08
90%	2	1	2	2	1	3	3	4	2	0	4	3	2	1	3	1	32	1,28
Total	6	7	9	6	5	6	5	6	4	3	6	3	7	2	5	2	77	3,08
%	7,79%	9,09%	11,69%	7,79%	6,49%	7,79%	6,49%	7,79%	3,90%	3,90%	7,79%	3,90%	9,09%	2,60%	6,49%	2,60%		

Table 5: Overview of Empirical Analysis per Industry Sector

sector. This observation is also important for the consumer sector because the same conditions apply there. In the energy sector, the total number of findings is under the level of the technology and consumer sector, but there patents are also important in the financial communication.

5.4.3 Results of Empirical Analysis per Publication Years

The investigation period is from 2003 to 2012. Based on Table 6 in all years, there are significant patents. The largest number of patents is in 2008 with 27 out of 33 significant patents. That is 82% of all patents. The lowest number of patents is in 2003 with six out of 15 patents or 40% patents of all patents. In 2012, the number of significant patents is 13 out of 31 patents or 42%. Apart of 2004 with 11 out of 22 patents or 50%, the range of significant patents is between 60% and 80%.

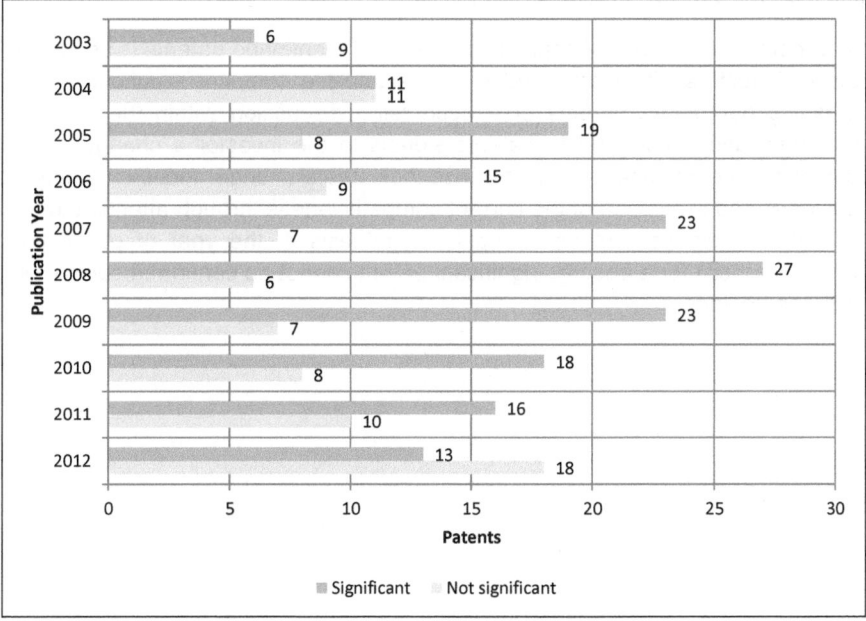

Figure 14: Significance per Publication Year
Own diagram according to: Appendix 1.

The constant range of significant patents in the whole investigation period confirms the general influence of intellectual property on the share price performance. But a general statement due to favorable years is not possible. There is dependency between general economic development and development in the field of R&D. During good economic years, the number of significant as well as the total number of patents is higher than during bad economic years.

The general importance of patents for the share price is additionally confirmed by the constant number of findings per each patent. There are between 1,7 and 3,0 findings during the whole investigation period. There is just one exception in 2005. There are 4,21 findings per patent in general.

These findings support the findings of Tikoo and Ebrhim. Companies are engaged in R&D during good economic times. In bad economic times, companies focus on things pushing the operative business like marketing. Therefore, companies invest during good economic times in long-term R&D, etc., which assure their position within the competition.[184]

In general, the number of applied patents or all mentioned patents is an indicator for investors. They are needed to evaluate the company's outlook. A decreasing number of patents or a decreasing discussion within the financial communication according to patents signals to the investor a change in the company's development. The decrease indicates a stronger focus on safety in business and protection of the current competitiveness through attention to the operative business. But according to Hypothesis 4, the year of publication which is crucial for a positive significant patent cannot be confirmed completely with the findings.

---

[184]   Cp. Tikoo, S., Ebrahim, A. (2010), p. 55.

Table 6: Overview of Empirical Analysis per Publication Year

5.4.4 Results of Empirical Analysis per Sales

As Table 7 shows, the data set includes values from different companies with different amounts of sales. Sales are in the range of less than 10 Mio. EUR up to 50 Mio. EUR per year. In the analysis, most companies with 118 belong to the category less than 10 Mio. EUR sales per year.

There are in total five different categories. The first category covers all companies with an amount of total sales less than 10 Mio. EUR. In this category, there are 118 patents. Out of 118 patents 81 are significant, which is 69%. In the second category, 27 out of 38 patents (71%) are significant. The third category has the lowest number of patents with 22 out of 39 patents (56%). In the fourth category, 29 out of 48 patents (60%) are significant. The last category includes 22 patents in total. Out of these 14 patents are significant. In all, the range of significant patents is between 56% and 71%.

Considering the previous points, there is no general statement regarding the relation between sales and patents. Hypothesis 5: The amount of company's sales is crucial for a positive significant patent cannot be confirmed. In all categories, more than half the patents are significant. This is just a general indicator that patents are important for SMEs' share price performance. There is, however, no special statement that a specific amount of sales has disproportionate effects on share price performance. The number of significant periods within each class confirms the previous mentions. The range of significant periods within the categories is from 2,38 significant findings up to 3,14 findings per patent. The small deviation confirms the general importance of patents, but it also confirms that there is no direct relation between the amount of sales and share price performance.

According to previous mentions, also in Table 7, it becomes apparent that there are more significant findings in the run-up to the publication date than in the follow-up. This can be an indication that there are investors who get to know several days prior to the publication. This observation confirms the point that small and medium-sized enterprises need a good stakeholder management. The number of stakeholders in SMEs is smaller than in big companies.[185] Investors are very important stakeholders of a company. The relationship between a company and its investors is closer in an SME. In contrast to big companies, the company knows a lot of its investors. The findings suggest that investors are closely related to the company invest in the run-up to receive a benefit from investment in the long-run. In this way, acting investors include private equity investors or other investors with a big involvement in a company.

---

[185] Cp. Krüger, W. (2006), pp. 19-21.

	<10	<20	<30	<40	>40	t-5	t-4	t-3	t-2	t-1	t	t+1	t+2	t+3	t+4	t+6	[-6; 6]	[-3; 0]	[-1; 0]	[0; 1]	[0; 3]	Total Significance	Significant Periods
Significant	81	27	22	29	14	39	38	31	38	31	28	38	28	31	28	37	15	22	22	21	20	467	
Total	118	38	39	48	22																		
Percentage	69%	71%	56%	60%	64%																		
99%	81					6	5	4	10	2	3	6	2	2	4	3	1	2	2	4	2	58	0,72
95%	81					7	6	7	5	4	6	4	3	4	7	7	1	3	1	4	3	72	0,89
90%	81					6	6	4	4	7	3	8	4	7	6	7	2	3	7	1	2	77	0,95
	19					19	17	15	19	13	12	18	9	13	17	17	4	8	10	9	7	207	2,56
	9,18%					9,18%	8,21%	7,25%	9,18%	6,28%	5,80%	8,70%	4,35%	6,28%	8,21%	8,21%	1,93%	3,86%	4,83%	4,35%	3,38%		
99%		27				3	3	3	2	1	0	2	3	3	2	0	2	1	1	1	3	31	1,15
95%		27				1	1	2	2	1	2	2	1	2	0	0	1	3	1	1	0	17	0,63
90%		27				2	1	0	3	2	4	3	4	2	2	4	1	1	1	1	1	31	1,15
		6				6	5	6	6	4	4	7	8	7	4	6	5	6	2	3	4	79	2,93
		7,59%				7,59%	6,33%	7,59%	6,33%	5,06%	5,06%	8,86%	10,13%	8,86%	5,06%	6,33%	5,06%	6,33%	2,53%	3,80%	5,06%		
99%			22			3	2	3	3	4	3	1	1	1	0	0	1	0	0	0	3	20	0,91
95%			22			3	3	2	3	0	1	1	1	0	1	0	2	2	0	1	0	22	1,00
90%			22			2	1	1	0	1	1	2	3	2	0	2	3	3	3	3	1	27	1,23
			6			6	6	7	6	5	5	4	5	3	1	5	5	5	3	4	2	69	3,14
			8,70%			8,70%	8,70%	10,14%	8,70%	7,25%	7,25%	5,80%	7,25%	4,35%	1,45%	7,25%	4,35%	7,25%	5,80%	4,35%			
99%				29		1	1	0	2	1	0	1	0	3	2	3	1	0	0	0	0	14	0,48
95%				29		2	3	3	0	2	3	3	2	0	1	4	1	1	2	2	0	26	0,90
90%				29		1	3	1	1	4	2	2	1	1	0	1	1	1	2	1	5	29	1,00
				4		4	7	1	6	7	6	6	3	5	3	8	2	2	3	3	6	69	2,38
				5,80%		5,80%	10,14%	1,45%	8,70%	10,14%	5,80%	8,70%	4,35%	7,25%	4,35%	11,59%	2,90%	2,90%	4,35%	4,35%	7,25%		
99%					14	0	1	2	1	0	0	1	1	1	0	2	0	1	0	2	0	7	0,50
95%					14	4	2	1	0	1	3	2	2	1	2	0	0	0	1	2	2	23	1,64
90%					14	0	0	1	2	1	0	0	1	1	2	0	2	2	1	2	0	13	0,93
					4	4	3	3	3	2	3	3	4	3	2	2	2	2	2	2	2	43	3,07
					9,30%	9,30%	6,98%	6,98%	6,98%	4,65%	6,98%	6,98%	6,98%	6,98%	4,65%	4,65%	4,65%	4,65%	4,65%	4,65%	4,65%		

Table 7: Overview of Empirical Analysis per Sale

5.4.5   Results of Empirical Analysis per Employees

The whole data set is split into five categories for analyzing the relation between patent publication and the number of employees in an SME. Regarding Table 8, the first category includes companies with less than 50 employees, category two covers all companies with less than 100 employees, in the third category there are companies with less than 150 employees, in category four companies have less than 200 employees and in category five less than 250 employees.

In category one, 22 out of 33 patents (67%) are significant. In category two, 59 out of 86 patents (69%) and 55 out of 77 patents (71%) in category three are significant. In categories four and five, the number of significant patents is lower with 12 out of 22 (55%) in category four and with 25 out of 47 (53%) in category five. Most significant patents are in total in the categories with the number of employees up to 150.

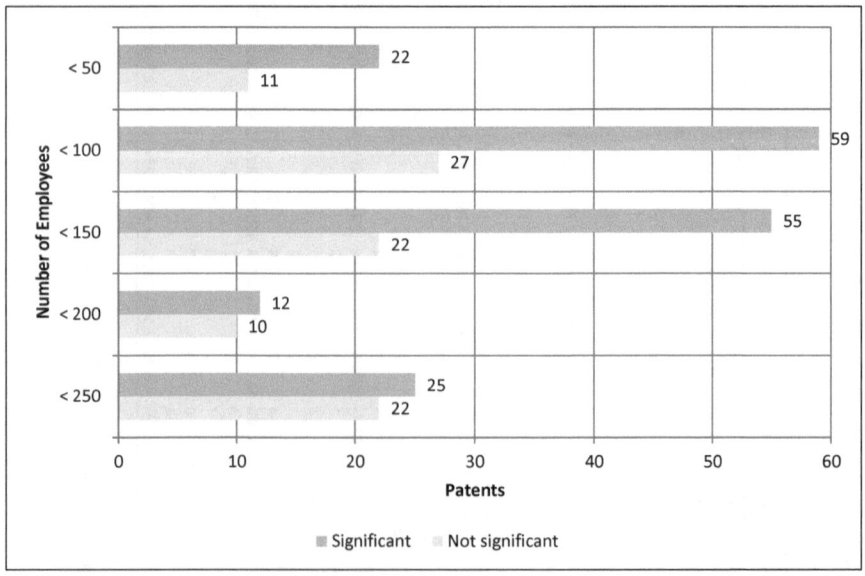

Figure 15: Significance per Employees
Own diagram according to: Appendix 1.

These findings confirm that innovations are very important for companies with a small number of employees. Thus, Hypothesis 6: The number of employees is crucial for confirming a positive significant patent. In such companies, all employees or huge number of employees are normally engaged in the process of patent application. The employees are proud of the innovation and their contribution to the success. They want to use the innovation on their own to be successful in business. In bigger companies, innovations are developed by a small number of employees. A lot of innovations do not fit the company's business. Therefore, big companies often sell their innovations in terms of licenses.[186] Companies with a small number of employees can benefit from patents in financial communication because patents in communication stress their attractiveness as well as the value of the human capital within the company. As Table 8 shows, the number of significant patents within one category supports the previously described findings.

Moreover, the point described several times that more significant patents in the run-up to the publication date is confirmed as well. A specific conclusion regarding the number of employees cannot be drawn in this case. There is just the general point that in total most significant patents are in the run-up.

---

[186]   Cp. Schwarz, E., Harms, R. (2005), p. 134.

	<50	<100	<150	<200	<250	t-5	t-4	t-3	t-2	t-1	t	t+1	t+2	t+3	t+4	t+5	[-5;5]	[-3;0]	[-1;0]	[0;1]	[0;3]	Total Significance	Significant Periods
Significant	22	59	55	12	25	39	38	32	38	31	28	38	28	31	28	37	15	22	22	21	20		468
Total	33	86	77	22	47																		
Percentage	67%	69%	71%	55%	53%																		
99%	22					1	0	1	1	1	0	2	0	0	2	0	0	1	1	1	0	11	0,50
95%	22					0	1	2	1	2	2	2	1	0	2	2	0	0	0	2	0	17	0,77
90%	22					1	3	0	2	2	3	2	1	1	1	2	1	1	3	0	2	24	1,09
Total	**2**					**2**	**4**	**3**	**4**	**5**	**5**	**6**	**2**	**1**	**5**	**4**	**1**	**1**	**4**	**3**	**2**	**52**	**2,36**
	3,85%					3,85%	7,69%	5,77%	7,69%	9,62%	9,62%	11,54%	3,85%	1,92%	9,62%	7,69%	1,92%	1,92%	7,69%	5,77%	3,85%		
99%		59				5	5	3	9	1	3	4	2	2	2	3	1	1	1	3	2	47	0,80
95%		59				7	5	5	4	2	4	2	2	2	5	5	3	1	1	2	3	55	0,93
90%		59				5	3	4	2	5	0	6	3	6	5	5	1	3	4	1	0	53	0,90
Total						**17**	**13**	**12**	**16**	**8**	**7**	**12**	**7**	**12**	**12**	**13**	**3**	**7**	**6**	**6**	**6**	**165**	**2,63**
						10,97%	8,39%	7,74%	9,68%	5,16%	4,52%	7,74%	4,52%	7,74%	7,74%	8,39%	1,94%	4,52%	3,87%	3,87%	3,23%		
99%			55			4	5	3	5	3	1	4	3	6	4	5	2	1	2	1	3	52	0,95
95%			55			6	4	3	2	2	6	6	4	3	1	3	2	2	5	2	2	55	1,00
90%			55			2	4	0	7	5	4	4	5	4	3	5	4	4	3	3	5	63	1,15
Total						**12**	**13**	**6**	**14**	**10**	**11**	**14**	**12**	**13**	**8**	**13**	**8**	**7**	**10**	**8**	**10**	**170**	**3,09**
						7,06%	7,65%	3,53%	8,24%	5,88%	6,47%	8,24%	7,06%	7,65%	4,71%	7,65%	4,71%	4,71%	5,29%	4,71%	5,88%		
99%				12		0	0	1	0	0	0	0	0	0	0	2	0	0	0	0	0	4	0,33
95%				12		0	1	0	0	2	1	1	1	1	1	1	0	1	0	1	0	9	0,75
90%				12		1	0	2	1	1	0	1	0	1	0	0	0	0	0	0	1	9	0,75
Total						**1**	**1**	**3**	**1**	**3**	**0**	**2**	**1**	**3**	**2**	**3**	**0**	**1**	**0**	**0**	**1**	**22**	**1,83**
						4,55%	4,55%	13,64%	4,55%	13,64%	0,00%	9,09%	4,55%	13,64%	9,09%	13,64%	0,00%	4,55%	0,00%	0,00%	4,55%		
99%					25	1	2	3	3	1	3	1	1	0	0	2	0	2	1	1	0	16	0,64
95%					25	4	4	3	3	0	1	1	1	0	1	2	2	1	2	1	1	25	1,00
90%					25	2	1	2	0	2	5	2	4	2	0	3	3	2	1	2	1	28	1,12
Total						**7**	**7**	**8**	**6**	**5**	**11**	**6**	**6**	**2**	**2**	**7**	**5**	**5**	**4**	**4**	**2**	**69**	**2,76**
						10,14%	10,14%	11,59%	6,80%	7,25%	7,25%	6,80%	8,70%	2,90%	1,45%	6,80%	4,35%	5,80%	6,80%	4,35%	2,90%		

Table 8: Overview of Empirical Analysis per Employees

6 Conclusion and Outlook

The general question of this work was to find out whether there is a significant relation between patent publication and share's performance in small and medium-sized companies. With regard to this general question, it is possible to draw a clear conclusion. Patents have a significant impact on the short-term share price performance of SMEs.

According to the theoretical background, patents are important for minimizing the information asymmetries between investor and company. The company is able to signal its capability to challenge business in the future. Patents are an orientation for company's position in the competition in the future. This point is especially important for small and medium-sized enterprises. The lower capital base and other points specific to SMEs make these companies more prone to crisis. Therefore, small and medium-sized enterprises must constantly signal their strengths to beat the competition.

Inventions are often innovations which have hidden values for companies. Externals such as investors do not get to know that there are such values. But these values are important for their investment decisions. Patents are written documents that stress the innovativeness of a company. An external higher institution confirms that the company has developed something new.

Based on the empirical findings, companies fulfilling the following parameters profit more than other companies from patents in their financial communication:
The findings within this work demonstrate the importance of patents in the financial communication of SMEs. 65% of all patents have a significant impact on the share price of a company. Particularly, for some companies, it is very important to put patents and other elements belonging to the field of intellectual properties in their financial communication strategies to position in the completion for capital and to enable good conditions for future refinancing:

1. Industry sector

Companies operating in the business areas with short product life-cycles and high innovative products, such as communication, with 77% significant patents, and the consumer sector, with 75% significant patents, make extraordinary profit from patents in financial communication. Investors look for innovative values within the companies. Through patents companies in the financial communication can improve the position in comparison with their competitors. This point is also valid for all sectors with a long product life-cycle and cost-intensive products like the energy sector. 80% of all patents from companies from the energy sector are significant.

2. Number of employees

In addition to the industry sector, the number of employees is important for the relationship between patents and share price performance. Companies with a small number of employees benefit from contents due to patents and other intellectual properties in their financial communication. They can improve their position in competition when they implement these contents in the financial communication. The deciding criterion in this case is that companies with a small number of employees mostly use the patent on their own and do not sell it as license to another player. Finally, 69% of all patents from companies with less than 150 employees have significant impacts on companies' share prices.

To summarize, it can be said that financial communication is very important for SMEs to make sure their future refinancing possibilities. Especially, investors investing in SMEs need to know the values of the company that strengthen the company's position with respect to other SMEs but also to big companies.

Owing to a growing capital market in Germany and the new bank lending conditions, SMEs have to improve their financial communication to ensure their future refinancing possibilities. Patents and other intellectual properties are good in this case because they demonstrate companies' capabilities to be competitive in the future. Competiveness is necessary to get investors who trust in the company's future. Trust is the basic requirement for getting capital from an external investor.

Bibliography

Acs, Z., Audretsch, D. (1990): Innovation and Small Firms, Cambridge 1990.

Agrawal, J., Kamakura, W. (1995): The Economic Worth of Celebrity Endorsers: An Event Study Analysis, in: Journal of Marketing, 1995, Iss. 59, pp. 56-62.

Ahlert, D., Schröder, H. (1996): Rechtliche Grundlagen des Marketing, 2. edn., Stuttgart 1996.

Ahsen, A., Heesen, M., Kuchenbuch, A. (2010): Grundlagen der Bewertung von Innovationen im Mittelstand, in: Ahsen, A. (ed.), Bewertung von Innovationen im Mittelstand, Berlin und Heidelberg 2010, pp. 1-38.

Albach, H. (1990): Innovation als Fetisch oder und Notwendigkeit, in: Albach, H. (ed.), Innovationsmanagement, Wiesbaden 1990, pp. 97-107.

Alchian, A., Demsetz, H. (1972): Production, Information Costs, and Economic Organization, in: American Economic Review, 1972, Vol. 62, pp. 777-795.

Alchian, A., Demsetz, H. (1973): The Property Rights Paradigm, in: Journal of Economic History, 1973, Vol. 33, pp. 16-27.

Alti, A. (2003): How sensitive is investment to cash flow when financing is frictionless?, in: Journal of Finance, 2003, Vol. 58, Iss. 2, pp. 707-722.

Arrow, K. (1969): The Organization of Economic Activity: Issues Pertinent to the Choice of Market versus Non-market Allocation, in: The Analysis and Evaluation of Public Expenditures: The PBBSystem, Joint Economic Committee, 91st Congress, 1st Session, Vol. 1, Washington, DC 1969.

Audretsch, D., Elston, J. (1997): Financing the German Mittelstand, in: Small Business Economics, 1997, Vol. 9, pp. 97-110.

Ayyagari, M., Beck, T., Demirgüç-Kunt, A. (2007): Small and medium enterprises across the globe, in: Small Business Economics, 2007, Vol. 29, pp. 415-434.

Bähr-Seppelfricke, U. (1999): Diffusion neuer Produkte, Wiesbaden 1999.

Ball, R., Brown, P. (1968): An empirical evaluation of accounting income numbers, in: Journal of Accounting Research, 1968, Vol. 6, pp. 159-178.

Bartscherer, M. (2004): Investor Relations in Versicherungsunternehmen (-konzernen), in: Koch, G., Köhne, T., Wagner, F. (eds.), Leipziger Schriften zur Versicherungswissenschaft, No. 6, Karlsruhe 2004.

Basel Committee on Banking Supervision (2000): Range of Practice in Banks' Internal Ratings Systems, in: Basel Committee Publications, No. 66, Basel 2000.

Bashmakov, I. (2009): Russian Energy Efficiency Potential: Scale, Costs, and Benefits, in: Problems of Economic Transition, 2009, Vol. 52, No. 1, pp. 54-75.

Beaver, W. (1968): The Information Content of Annual Earnings Announcements, in: Journal of Accounting Research, 1968, Vol. 6, pp. 67-92.

Becchetti, L., Castelli, A., Hasan, I. (2010): Investment-cash flow sensitives, credit rationing and financing constraints in small and medium-sized firms, in: Small Business Economics, 2010, Vol. 35, pp. 467-497.

Beck, T., Demirgüç-Kunt, A., Laeven, L., Maksimovic, V. (2006): The determinants of financing obstacles, in: Journal of International Money Finance, 2006, Vol. 25, pp. 932-952.

Beck, T., Demirgüç-Kunt, A., Maksimovic, V. (2005): Financial and legal constraints to firm growth: does firm size matter?, in Journal of Finance, 2005, Vol. 60, pp. 137-177.

Ben-Shahar, D., Feldman, D. (2003): Signaling-Screening Equilibrium in the Mortgage Market, in: Journal of Real Estate Finance and Economics, 2003, Vol. 26, Iss. 2/3, pp. 157-178.

Berger, A., Kayshap, A., Scalise, J. (1995): The transformation of the U.S. banking industry: what a long strange trip it's been, in: Brookings Paper Econ Active, 1995, Vol. 2, pp. 155-219.

Berger, A., Klapper, L., Udell, G. (2001): The ability of banks to lend to informationally opaque small businesses, in: Journal of Bank Finance, 2001, Vol. 25, pp. 2127-2167.

Berger, A., Udell, G. (1996): Universal banking and the future of small business lending, in: Saunders, A., Walter, I. (eds.), Financial system design: the case for universal banking, Burr Ridge 1996, pp. 559–627.

Besley, T., Ghatak, M. (2005): Incentives, risk and accountability in organizations, in: Hutter, B., Power, M. (eds.), Organizational Encounters With Risk, Cambridge 2005.

Bigus, J. (2006): Staging of venture financing, investor opportunism and patent law, in: Journal of Business Finance & Accounting, 2006, Vol. 33, Iss. 7-8, pp. 939-960.

BIS (2010): Basel III: A global regulatory framework for more resilient banks and banking systems, December 2010.

Bittelmeyer, C. (2007): Patente und Finanzierung am Kapitalmarkt, Wiesbaden 2007.

Bolton, P., Scharfstein, D. (1996): Optimal debt structure and the number of creditors, in: Journal of Political Economy, 1996, Vol. 104, Iss. 1, pp. 1-25.

Booth, G., Juttila, J., Kallunki, J., Rahiala, M. Sahlström, P. (2006): How Does the Financial Environment Affect the Stock Market Valuation of R&D Spending?, in: Journal of Financial Intermediation, 2006, Vol. 15, Iss. April, pp. 197-214.

Bound, J., Cummies, C., Griliches, Z., Hall, B., Jaffe, A. (1984): Who does R&D and who patents?, in: Griliches, Z. (ed.), R&D, Patents and Productivity, Chicago 1984.

Breid, V. (1995): Aussagefähigkeit agencytheoretischer Ansätze im Hinblick auf die Verhaltenssteuerung von Entscheidungsträgern, in: zfbf – Schmalenbachs Zeitschrift für betriebswirtschaftliche Forschung, 1995, Vol. 47, No. 9, pp. 821-854.

Brown, S., Warner, J. (1980): Measuring security price performance, in: Journal of Financial Economics, 1980, Vol. 8, Iss. 3, pp. 205-258.

Brown, S., Warner, J. (1985): Using daily stock returns: the case of event studies, in: Journal of Financial Economics, 1985, Vol. 14, Iss. 1, pp. 3-31.

Cable, J. (1985): Capital Market Information and Industrial Performance: The Role of West German Banks, in: Economic Journal, 1985, Vol. 95, pp. 118-132.

Carlin, W., Richthofen, P. (1995): Finance, Economic Development and the Transition: The East German Case, Wissenschaftszentrum für Sozialforschung, Discussion Paper FS I95-301, Berlin 1995.

Chan, H., Martin, J., Kensinger, J. (1990): Corporate Research and Development Expenditures and Share Value, in: Journal of Financial Economics, 1990, Vol. 26, Iss. 2, pp. 255-276.

Chauvin, K., Hirschey, M. (1993): Advertising, R&D Expenditures and the Market Value of the Firm, in: Financial Management, 1993, Vol. 22, Iss. 4, pp. 128-141.

Cheng, F., Ariff, M. (2007): Abnormal Returns of Bank Stocks and Their Factor-Analyzed Determinants, in: Journal of Accounting – Business & Management, 2007, Vol. 14, pp. 1-16.

Chung, K., Wright, P., Kedia, B. (2003): Corporate Governance and Market Valuation of Capital and R&D Investments, in: Review of Financial Economics, 2003, Vol. 12, Iss. 2, pp. 161-172.

Clement, M., Fischer, M., Goerke, B. (2007): Neuprodukteinführungen in der Filmindustrie: Wie reagieren Kapitalmarktinvestoren auf den Umsatzerfolg neuer Kinofilme?, in: Die Betriebswirtschaft, 2007, Iss. 67, pp. 418-444.

Coase, R. (1960): The Problem of social cost, in: Journal of Law and Economics, 1960, Vol. 3, pp. 1-44.

Coase, R. (1984): The New Institutional Economics, in: Journal of Institutional and Theoretical Economics, 1984, Vol. 140, pp. 229-231.

Cohen, W. (1995): Empirical studies of innovative activity, in: Stoneman, P. (ed.), Handbook of Innovation and Technological Change, Oxford 1995.

Cohen, W., Levin, R. (1989): Empirical studies of innovation and market structure, in: Schmalensee, R., Willig, R. (eds.), Handbook of Industrial Organization, Vol. 2, Amsterdam 1989.

Combs, K. (1992): Cost sharing vs. multiple research projects in cooperative R&D, in: Economic Letters, 1992, Iss. 39, pp. 353-371.

Cong, Y., Hoitash, R., Krishman, M. (2010): Event Study with Imperfect Competition and Private Information: Earnings Announcements Revisited, in: Review of Quantitative Finance and Accounting, 2010, Vol. 34, Iss. 3, pp. 383-411.

Conlisk, J. (1996): Why Bounded Rationality?, in: Journal of Economic Literature, 1996, Vol. 34, pp. 669-700.

Corrado, C. (2011): Event studies: A methodology review, in: Accounting & Finance, 2011, Vol. 51, pp. 207-234.

d'Aspremont, C., Jacquemin, A (1988): Cooperative and non-cooperative R&D in duopoly with spillovers, in: American Economic Review, 1988, Iss. 78, pp. 1133-1137.

Davis, L., North, D. (1971): Institutional Change and American Economic Growth, Cambridge 1971.

Davis, S., Haltiwanger, J., Schuh, S. (1996): Small Business and Job Creation: Dissecting the Myth and Reassessing the Facts, in: Small Business Economics, 1996, Vol. 8, pp. 297-315.

Demsetz, H. (1967): Toward a Theory of Property Rights, in: American Economic Review, 1967, Vol. 57, Iss. 2, pp. 347-359.

Detragiache, E., Garella, P., Guiso, L. (2000): Multiple versus single banking relationships: Theory and evidence, in: Journal of Finance, 2000, Vol. 55, Iss. 3, pp. 1133-1161.

Deutsche Börse (2012): Prime Standard und General Standard – International anerkannte Marktsegmente: Zulassungsvoraussetzungen und wesentliche Folgepflichten, Frankfurt 2012.

Duca, J., Muellbauer, J., Murphy, A. (2010): Housing markets and the financial crisis of 2007–2009: lessons for the future, in: Journal of Financial Stability, 2010, Vol. 6, No. 4, pp. 203-217.

Edwards, D. (2001): Patent Backed Securitization: Blueprint For a New Asset Class, 2001.

Eisenhardt, K. (1989): Agency theory - An Assessment and Review, in: Academy of Management Review, 1989, Vol. 14, Iss. 1, pp. 57-74.

Ensminger, J. (1992): Making a Market, Cambridge 1992.

Ensthaler, J., Strübbe, K. (2006): Patentbewertung, Berlin und Heidelberg 2006.

EPÜ (2000): Europäisches Patentübereinkommen vom 05.10.1973 mit allen späteren Änderungen in der Fassung vom 29.11.2000.

Erlei, M., Leschke, M., Sauerland, D. (2007): Neue Institutionenökonomik, 2. edn., Stuttgart 2007.

European Commission (2005): The new SME definition, Brussels 2005.

Fama, E. (1980): Agency Problems and the Theory of the Firm, in: Journal of Political Economy, 1980, Vol. 88, Iss. 2, pp. 288-307.

Fama, E., Fisher, L., Jensen, M., Roll, R. (1969): The adjustment of stock prices to new information, in: International Economic Review, 1969, Vol. 10, No. 1, pp. 1-21.

Fama, E., Jensen, M. (1983a): Separation of Ownership and Control, in: Journal of Law and Economics, 1983, Vol. 26, pp. 301-325.

Fama, E., Jensen, M. (1983b): Agency Problems and Residual Claims, in: Journal of Law and Economics, 1983, Vol. 26, pp. 327-349.

Flueglistaller, U. (2004): Charakteristik und Entwicklung von Klein- und Mittelunternehmen (KMU), St. Gallen 2004.

Frank, R. (1994): Microeconomics and Behavior, 2. edn., New York 1994.

Furubotn, E. Richter, R. (2005): Institutions and Economic Theory. The Contribution of the New Institutional Economics, 2. edn., Michigan 2005.

Furubotn, E., Pejovich, S. (1972): Property Rights and Economic Theory: A Survey of Recent Literature, in: Journal of Economic Literature, 1972, Vol. 10, pp. 1137-1162.

Furubotn, E., Richter, R. (2008): The New Institutional Economics – A Different Approach To Economic Analysis, in: Economic Affairs, 2008, Vol. 28, Iss. 3, pp. 15-23.

Gambacorta, L., Marques-Ibanez, D. (2011): The Bank Lending Channel: Lessons from the Crisis, in: Economic Policy, 2011, Iss. 66, pp. 135-168, 175-182.

Gelbmann, U., Vorbach, S., Zotter, K. (2004): Konzepte für das Innovationsmanagement in Klein- und Mittelunternehmen, in: Schwarz, E. (ed.), Nachhaltiges Innovationsmanagement, Wiesbaden 2004, pp. 248-273.

Gerke, W. (2005): Kapitalmärkte – Funktionsweisen, Grenzen, Versagen, in: Hungenberg, H., Meffert, J. (eds.), Handbuch Strategisches Management, 2. edn., Wiesbaden 2005.

Gigerenzer, G., Selten, R. (2001): Bounded Rationality: The Adaptive Toolbox, Cambridge 2001.

Gogoris, A., Clarke, P. (2001): Patent due diligence in biotechnology transactions, in: Nature biotechnology, 2001, Vol. 19, Iss. 3, pp. 279-281.

Goto, A. (2009): Innovation and Competition Policy, in: The Japanese Economic Review, 2009, Vol. 60, No. 1, pp. 56-62.

Graham, S., Merges, R., Samuelson, P., Sichelman, T. (2009): High technology entrepreneurs and the patent system: Results of the 2008 Berkeley patent survey, in: Berkeley Technology Law Journal, 2009, Vol. 24, Iss. 4, pp. 255-327.

Graumann, S., Keil, M. (2004): Neue Methoden zur Messung der PR-Effizienz, dargestellt an einem Fallbeispiel aus der Praxis. Vergleich der öffentlichen versus der veröffentlichten Meinung, in: tns infratest Business Intelligence, München 2004.

Griliches, Z. (1990): Patent statistics as economic indicators: a survey, in: Journal of Economic Literature, 1990, Iss. 18, pp. 1661-1707.

Grossmann, S., Hart, O. (1986): The Costs and Benefits of Ownership: A Theory of Vertical and Lateral Integration, in: Journal of Political Economy, 1986, Vol. 94, Iss. 4, pp. 694-719.

Grunig, J., Hunt, T. (1984): Managing public relations, New York 1984.

Grüning, M. (2011): Publizität börsennotierter Unternehmen, Wiesbaden 2011.

Gzybowski, M. (2013): Recent Patent Activity, in: IPINDEPTH, 2013, Iss. June 2013, pp. 9-10.

Haugg, N., Lokys, A., Winterfeldt, V. (1989): Das national und internationale Patentsystem, in: Engelhardt, K. (ed.): Fachwissen Patentinformation: Datenbanken strategisch genutzt, Essen 1989, pp. 183-197.

Hauschildt, J., Salomo, S. (2007): Innovationsmanagement, 4. edn., München 2007.

Hayek, F. (1973): Law, Legislation, and Liberty: Rules and Order, Vol. 1, University of Chicago Press, Chicago 1973.

Herstatt, C., Lüthje, C., Verworn, B. (2001): Die Gestaltung von Innovationsprozessen in kleinen und mittleren Unternehmen, in: Meyer, J. (ed.), Innovationsmanagement in kleinen und mittleren Unternehmen, München 2001, pp. 149-169.

Hillmann, M. (2011): Unternehmenskommunikation kompakt – Das 1 x 1 für Profis, Wiesbaden 2011.

Hirschey, M., Richardson, V., Scholz, S. (2001): Value relevance of nonfinancial information: The case of patent data, in: Review of Quantitative Finance and Accounting, 2001, Vol. 17, Iss. 3, pp. 223-235.

Holmström, B. (1979): Moral hazard and observability, in: Journal of Economics, 1979, Vol. 10, Iss. 1, pp. 74-91.

Holmström, B. (1982): Moral hazard in teams, in: Journal of Economics, 1982, Vol. 13, Iss. 2, pp. 324-340.

Hsu, D., Ziedonis, R. (2008): PATENTS AS QUALITY SIGNALS FOR ENTREPRENEURIAL VENTURES, in: Academy of Management Proceedings, 2008, pp. 1-6.

Hubig, C., Siemoneit, O. (2009): Vertrauen und Glaubwürdigkeit als kommunikationspolitische Ziele erfolgreicher IR, in: Kirchhoff, K., Piwinger, M. (eds.), Praxishandbuch Investor Relations – Das Standardwerk der Finanzkommunikation, 2. edn., Wiesbaden 2009, pp. 63-72.

Huchzermeier, M. (2006): Investor Relations beim Börsengang – Konzept für mittelständische Unternehmen, Wiesbaden 2006.

IADB (2004): Unlocking credit: the quest for deep and stable lending, Baltimore 2004.

Jensen, M. (1983): Organization Theory and Methodology, in: Accounting Review, 1983, Vol. 56, Iss. 2, pp. 319-339.

Jensen, M., Meckling, W. (1976): The Theory of the Firm: Managerial Behavior, Agency Costs and Ownership Structure, in: Journal of Financial Economics, 1976, Vol. 3, Iss. 4, pp. 305-360.

Kamien, M., Muller, E., Zang, I. (1992): Research joint ventures and R&D cartels, in: American Economic Review, 1992, Iss. 82, pp. 1293-1306.

Keeton, W. (1995): Multi-office bank lending to small businesses: some new evidence, in: Fed Reserve Bank Kansas City Economic Review, 1995, Vol. 80, pp. 45–57.

Kester, W. (1992): Industrial Groups as Systems of Contractual Governance, in: Oxford Review of Economic Policy, 1992, Vol. 8, Iss. 3, pp. 24-44.

Kim, J., Mahony, J. (2005): Property Rights Theory, Transaction Costs Theory, and Agency Theory: An Organizational Economics Approach to Strategic Management, in: Managerial and Decision Economics, 2005, Vol. 26, pp. 223-242.

Klein, B., Crawford, R., Alchian, A. (1978): Vertical Integration, Appropriate Rents, and the Competitive Contracting Process, in: Journal of Law and Economics, 1978, Vol. 21, Iss. 2, pp. 279-326.

Klette, T., Griliches, Z. (2000): Empirical patterns of firm growth and R&D investment: A quality ladder model interpretation, in: The economic journal, 2000, Vol. 110, Iss. April, pp. 363-387.

Kolari, J., Pynnönen, S. (2010): Event Study Testing with Cross-sectional Correlation of Abnormal Returns, in: in: The Review of Financial Studies, 2010, Vol. 23, No. 11, pp. 3996-4025.

Kothari, S., Warner, J. (2007): Econometrics of event studies, in: Eckbo, B., ed., Handbook of Corporate Finance: Empirical Corporate Finance, Amsterdam and Oxford 2007, pp. 3-36.

Krämer, W. (2003): Mittelstandsökonomik, München 2003.

Krüger, W. (2006): Standortbestimmung – Wo steht der Mittelstand?, in: Krüger, W., Klippstein, G., Merk, R., Wittberg, V. (ed.), Praxishandbuch Mittelstand, Wiesbaden 2006.

Laffont, J., Martimort, D. (2002): The Theory of Incentives – The Principal-Agent Model, New Jersey and Woodstock 2002.

Larcker, D., Gordan, L., Pinches, G. (2001): Testing for market efficiency: a comparison of the cumulative average residual methodology and intervention analysis, in: Journal of Financial and Quantitative Analysis, 2001, Vol. 2, pp. 267-287.

Laskin, A. (2006): Investor relations practices at Fortune-500 companies: An exploratory study, in: Public Relations Review, 2006, Vol. 32, pp. 69-70.

Lehmann, E., Neuberger, D., Räthke, S. (2004): Lending to Small and Medium-Sized Firms: Is There an East-West Gap in Germany?, in: Small Business Economics, 2004, Vol. 23, pp. 23-29.

Macharzina, K., Wolf, J. (2008): Unternehmensführung, 6. edn., Wiesbaden 2008.

Marjit, S. (1991): Incentives for cooperative and non-cooperative R&D in duopoly, in: Economic Letters, 1991, Iss. 37, pp. 187-191.

McWilliams, A., Siegel, D. (1997): Event Studies in Management Research: Theoretical and Empirical Issues, in: Academy of Management Journal, 1997, Iss. 40, pp. 626-657.

Menninger, J., Kunowski, S. (2003): Wertermittlung von Patenten, Technologien und Lizenzen vor dem Hintergrund von Optimierungsstrategien, München 2003.

Mian, A. (2006): Distance constraints: the limits of foreign lending in poor economies, in: Journal of Finance, 2006, Vol. 61, Iss. 3, pp. 1465-1505.

Miller, M., Modigliani, F. (1961): Dividend policy, growth and the valuation of shares, in: Journal of Business, 1961, Vol. 34, pp. 411-433.

Miller, M., Modigliani, F. (1963): Corporate income taxes and the cost of capital: a correction, in: American Economic Review, 1963, Vol. 53, Iss. 3, pp. 433-443.

Minder, S. (2001): Wissensmanagement in KMU, St. Gallen 2001.

Modigliani, F., Miller, M. (1958): The cost of capital, corporation finance and the theory of investment, in: American Economic Review, 1958, Vol. 48, Iss. 3, pp. 261-297.

Mugler, J. (2005): Grundlagen der BWL der Klein- und Mittelbetriebe, Wien 2005.

Mukherjee, A., Marjit, S. (2004): R&D organization and technology transfer, in: Group Decision and Negotiation, 2004, Iss. 13, pp. 243-258.

North, D. (1986): The New Institutional Economics, in: Journal of Institutional and Theoretical Economics, 1986, Vol. 142, pp. 230-237.

North, D. (1990): Institutions, Institutional Change and Economic Performance, Cambridge 1990.

North, D. (1995): The New Institutional Economics and Third World Development, in: Harriss, J., Hunter J., Lewis, C. (eds.), The New Institutional Economics and Third World Development, London 1995, pp. 17-26.

Osterrieth, C. (2004): Patentrecht, 2. edn., München 2004.

Pakes, A., Schankerman, M. (1984): An exploration into the determinants of research intensity, in: Griliches, Z. (ed.), R&D, Patents and Productivity, Chicago 1984.

PatG (2011): Patentgesetz from 16.12.1980 (BGBl. 1981 I p. 1) with all later changes in the version from 24.11.2011. In: BGBl. I p. 2302.

Pedersen, A. (2008): Investor Relations Newsletters – Funktion, Aufbau und Sprache, in: Siems, F., Brandstätter, M., Gölzner, H. (eds.), Anspruchsgruppenorientierte Kommunikation – Neue Ansätze zu Kunden-, Mitarbeiter- und Unternehmenskommunikation, Wiesbaden 2008, pp. 371-385.

Perl, E. (2007): Grundlagen des Innovations- und Technologiemanagements, in: Strebel, H. (ed.), Innovations- und Technologiemanagement, 2. edn., Wien 2007, pp. 17-52.

Petersen, M., Rajan, R. (1994): The benefits of lending relationships: Evidence from small business data, in: Journal of Finance, 1994, Vol. 49, pp. 1367–1400.

Pfohl, H. (2006): Abgrenzung der Klein- und Mittelbetriebe von Großbetrieben, in: Pfohl, H. (ed.), Betriebswirtschaftslehre der Mittel- und Kleinbetriebe, 4. edn., Berlin 2006, pp. 2-24.

Picot, A., Dietl, H., Franck, E. (2008): Organisation – Eine ökonomische Perspektive, 5. edn., Stuttgart 2008.

Picot, A., Reichwald, R., Wigand, R. (2003): Die grenzenlose Unternehmung: Information, Organisation und Management, 5. edn., Wiesbaden 2003.

Pinches, G., Narayanan, V., Kelm, K. (1996): How the Market Values the Different Stages of Corporate R&D - Initiation, Progess, and Commercialization, in: Journal of Applied Corporate Finance, 1996, Vol. 9, Iss. 1, pp. 60-69.

Pindado, J., Rodrigues, L. (2004): Parsimonious Models of Financial Insolvency in Small Companies, in: Small Business Economics, 2004, No. 22, pp. 51-66.

Piwinger, M. (2009): IR als Kommunikationsdisziplin, in: Kirchhoff, K., Piwinger, M. (eds.), Praxishandbuch Investor Relations – Das Standardwerk der Finanzkommunikation, 2. edn., Wiesbaden 2009, pp. 13-34.

Pleschak, F., Sabisch, H. (1996): Innovationsmanagement, Stuttgart 1996.

Rebel, D. (2001): Gewerbliche Schutzrechte. Anmeldung – Strategie – Verwertung, 3. edn., Köln 2001.

Reitzig, M. (2002): Die Bewertung von Patentrechten: Eine theoretische und empirische Analyse aus Unternehmenssicht, Wiesbaden 2002.

Restaino, L. (2006): Understanding Your Patent Portfolio: Reducing Risk Through Due Diligence, in: BioProcess International, 2006, Iss. October, pp. 12-16.

Rogers, E. (2003): Diffusion of innovations, 3. edn., London 2003.

Ruppert, D. (2004): Statistics and Finance – An Introduction, New York 2004.

Schmoch, U. (1990): Wettbewerbsvorsprung durch Patentinformation: Handbuch für die Recherchenpraxis, Köln 1990.

Schulte, R., Tegtmeier, S., Eggers, F., Deutschmann, M. (2005): Von der Innovation zum Corporate Spin Off – Technologieorientierte Ausgründungen aus Großunternehmen in Deutschland, in: Schwarz, E., Harms, E. (ed.), Integriertes Ideenmanagement: betriebliche und überbetriebliche Aspekte unter besonderer Berücksichtigung kleiner und junger Unternehmen, Wiesbaden 2005.

Schumpeter, J. (1942): Capitalism, Socialism and Democracy, London 1942.

Sengupta, R. (2007): Foreign entry and bank competition, in: Journal of Financial Economics, 2007, Vol. 84, Iss. 2, pp. 502-528.

Sharpe, W. (1963): A SIMPLIFIED MODEL FOR PORTFOLIO ANALYSIS, in: Management Science, 1963, Vol. 9, Iss. 2, pp. 277-293.

Sharpe, W. (1964): Capital asset prices: a theory of equilibrium under conditions of risk, in: Journal of Finance, Vol. 19, pp. 425-442.

Shevlin, T. (1981): Measuring abnormal performance on the Australian securities market, in: Australian Journal of Management, 1981, Vol. 6, Iss. 1, pp. 67-107.

Smith, A. (1776): An Inquiry into the Nature and the Causes of the Wealth of Nations, 1776. (Republished 2008 by Forgotten Books New York).

Specht, G., Beckmann, C., Amelingmeyer, J. (2002): F&E-management, 2nd edn., Stuttgart 2002.

Specht, G., Beckmann, C., Amelingmeyer, J. (2002): F&EManagement: Kompetenz im Innovationsmanagement, 2. edn. Stuttgart 2002.

Spence, M. (1973): Job market signaling, in: Quarterly Journal of Economics, 1973, Vol. 87, Iss. 3, pp. 355-374.

Stanzel, M. (2007): Determinanten der Researchqualität von Finanzanalysten, Wiesbaden 2007.

Strahan, P., Weston, J. (1996): Small business lending and bank consolidation: is there cause for concern? Current Issues in Economics and Finance, in: Federal Reserve Bank of New York, 1996, Vol. 2, pp. 1-6.

Strong, N. (1992): Modelling abnormal returns: A preview article, in: Journal of Business Finance & Accounting, 1992, Vol. 19, Iss. 4, pp. 533-553.

Suzumura, K. (1992): Cooperative and non-cooperative R&D in an oligopoly with spillovers, in: American Economic Review, 1992, Iss. 82, pp. 1307-1320.

Thadden, E. (1995): Long term contracts, short term investment and monitoring, in: Review of Economic Studies, 1995, Vol. 62, pp. 557-575.

Tikoo, S., Ebrahim, A. (2010): Financial Markets and Marketing: The Tradeoff between R&D and Advertising during an Economic Downturn, in: Journal of Advertising Research, 2010, Vol. 50, No. 1, pp. 5056.

Ueda, M. (2004): Banks versus Venture Capital: Project Evaluation, Screening, and Expropriation, in: Journal of Finance, 2004, Vol. 59, Iss. 2, pp. 601-625.

Vahs, D., Burmester, R. (2005): Innovationsmanagement, 3. edn., Stuttgart 2005.

Walther, S. (2004): Erfolgsfaktoren von Innovationen in mittelständischen Unternehmen, Frankfurt am Main 2004.

Wehinger, G. (2012): Bank deleveraging, the move from bank to market-based financing, and SME financing, in: OECD Journal: Financial Market Trends, 2012, Vol. 1, pp. 66-79.

Williamson, O. (1967): Hierarchical Control and Optimum Firm Size, in: Journal of Political Economy, 1967, Vol. 75, pp. 123-138.

Williamson, O. (1975): Markets and hierarchies, analysis and antitrust implications: A study in the economics of internal organization, New York 1975.

Williamson, O. (1981): The Modern Corporation: Origin, Evolution, Attributes, in: Journal of Economic Literature, 1981, Vol. 19, pp. 1537-1568.

Williamson, O. (2000): The New Institutional Economics: Taking Stock, Looking Ahead, in: Journal of Economic Literature, 2000, Vol. 38, pp. 595-613.

Wurzer, A. (2004): Patentmanagement: ein Praxisleitfaden für den Mittelstand, Eschborn 2004.

Appendix 1: Empirical Analysis

To access the book's appendix, please visit www.springer.com and search for the author's name.

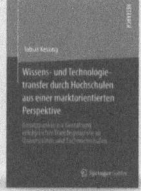